T-Labs Series in Telecommunication Services

Series editors

Sebastian Möller, Quality and Usability Lab, Technische Universität Berlin, Berlin, Germany

Axel Küpper, Telekom Innovation Laboratories, Technische Universität Berlin, Berlin, Germany

Alexander Raake, Audiovisual Technology Group, Technische Universität Ilmenau, Ilmenau, Germany

More information about this series at http://www.springer.com/series/10013

Benjamin Weiss

Talker Quality in Human and Machine Interaction

Modeling the Listener's Perspective
in Passive and Interactive Scenarios

 Springer

Benjamin Weiss
Technische Universität Berlin
Berlin, Germany

ISSN 2192-2810   ISSN 2192-2829   (electronic)
T-Labs Series in Telecommunication Services
ISBN 978-3-030-22771-5   ISBN 978-3-030-22769-2   (eBook)
https://doi.org/10.1007/978-3-030-22769-2

This Springer imprint is published by the registered company Springer Nature Switzerland AG.
The registered company address is: Gewerbestrasse 11, 6330 Cham, Switzerland

To Ulrike and Toni

Preface

When humans engage in spoken interaction, they assess the obvious features of the interaction partner, such as the verbal and nonverbal signals of the voice, as these contribute to the subjective evaluation of the interlocutor. In this book, the background, state of research, and own contributions to the **assessment and prediction of talker quality** that is constituted in voice perception and in dialog are presented. Starting from theories and empirical findings from human interaction, major results and approaches are transferred to the domain of human-computer interaction. The main subject of this book is to contribute to the evaluation of spoken interaction in both humans and between human and computer, and in particular to the quality subsequently attributed to the speaking system or person, based on the listening and interactive experience.

The theories, methods, and results presented are focused on the **first impression** of the people engaged in such a vocal conversation. This means hearing a voice for the first time in either a passive scenario (listening only) or interacting with a person or computer for the first time (interactive scenario). By using the term "first impression," the research focus is set to the human perception and evaluation of voices and conversations experienced, which represents the beginning attitude formation of the participating human towards the (other) speaker, may it be a real person or a computer agent.

The main scientific contribution is not to psychological theory development, but it is an informed engineering approach for describing subjective quality as experienced by users. As mentioned in the subtitle, the major results of this book are the development of **quantitative models of user ratings** to represent subjective quality. The most important part of this modeling is the **identification of relevant parameters** as predictors. Such predictors of talker quality can be acoustic features, for the case of voices' quality that is typically assessed in passive scenarios, or data describing interaction behavior for the course of conversation. By modeling voice-stimulated user ratings, this book also contributes to the identification of the most important **perceptual dimensions**—as these factors provide the valuable insight in search for quantitative predictors, and derives basic insights and principles in

designing and evaluating modern spoken conversational systems, with the aim of increasing and ensuring quality.

This book is intended for advanced readers, who have already a background in speech signal processing and some basic knowledge in social psychology. Results on the topic of talker appraisal are presented mostly in a rather condensed form, and not all basic terms are defined or explained. Several of my own contributions, in addition to other work and results presented here, are of course published elsewhere. Therefore, please use the references to obtain more details and additional information if the respective summarized section does not answer all your questions.

Berlin, Germany   Benjamin Weiss
February 2019

Acknowledgments

I want to thank first and foremost my mentor, Sebastian Möller, who supported me for over 10 years, in which I have worked on my favorite topics. He has built up a great team at the Quality and Usability Lab. There are so many current and former colleagues, too many to name here, without whom this book would not exist. Many are listed in the references, so I mention just Christine, Klaus, Ina, Felix, Irene, Yasmin, Stefan, Tilo, Thilo, Matthias, and Laura. On many conferences and on other occasions, I received valuable comments from colleagues in the field, in anonymous reviews but also in person while enjoying lively discussions. In particular, I want to thank Jürgen and Timo. My gratitude also goes to Petra Wagner and Elmar Nöth for writing reports on this manuscript for the habilitation committee. And lastly, I want to thank my family for being there for me.

I would like to acknowledge the financial support of the Deutsche Forschungs-gemeinschaft throughout my academic career, lastly by the projects "Sympathie von Stimme und Sprechweise Analyse und Modellierung auditiver und akustischer Merkmale" and "Human Perception and Automatic Detection of Speaker Personality and Likability: Influence of Modern Telecommunication Channels."

Contents

Chapter 1
Theory: Foundations of Quality in Natural and Synthesized Speech

Speech is one of the most important modes to communicate and interact in human–human interaction (HHI). It contains semantic and pragmatic meaning, often in an underspecified and indirect way, by referencing to situational and world knowledge. Apart from that, however, each utterance also includes nonverbal information, simply due to the prosody inevitably produced by speaking. Prosody is of great importance, as its variation signals linguistic and non-linguistic information, such as emphasis or syntactic and semantic structure, turn organization, or affective states. It comprises the primary [37] acoustic features of fundamental frequency, duration, and intensity, which are determining human perception of, e.g., intonation, speech rate, and loudness. Secondary prosodic features are more complex to be described acoustically, as they are caused by laryngeal and supra-laryngeal settings [213], such as voice quality or nasality, and by vocal tract shape or articulatory precision—all with multiple effects on acoustics. In order to support an easy understanding of such terms throughout the book, *speaking style* is used to cover intonation, pronunciation, and tempo, including pausing, whereas timbre and pitch level is comprised with the term *voice*. Primary and secondary prosodic features contribute to either of these two. Those aspects of the voice that are produced directly at the vocal folds will be called *voice quality*.

According to Schulz von Thun [319], the factual information of a message is accompanied in varying degrees by self-revelation (such as age, gender, emotion), relationship (e.g., social status, attitudes towards each other), and appeal (intention). All these four aspects are of course signaled multimodally and in a coordinated manner in HHI, e.g., by temporal relations between gestural, facial, and verbal signals [193], but this book focuses on speech. Therefore, most empirical work, including own contributions, is speech-only or within the domain of speech-based multimodal interaction.

Although some of these paralinguistic aspects—so called, as they traditionally did not contribute directly to the scientific core areas of linguistics, such as syntax

© Springer Nature Switzerland AG 2020
B. Weiss, *Talker Quality in Human and Machine Interaction*, T-Labs Series in Telecommunication Services, https://doi.org/10.1007/978-3-030-22769-2_1

or semantics—are intentionally signaled (e.g., friendliness by smiling), the majority can be considered as being automatically produced and therefore unavoidable (especially physiological information such as biological age group and sex, but also learned idiosyncrasies signaling a person's identity). In human–computer interaction (HCI), such implicit and unintentional signals (paralinguistics, gaze, gesturing) represent a potentially rich resource of information to be exploited and are therefore categorized as passive input mode [271], contrasting the active, i.e., explicit and intentional, mode of deliberately sending a message, e.g., for controlling a device.

Both, intentional and unintentional signals contribute to the important role paralinguistics have for a person's experience when listening to speech or engaging in a conversation. Whether this experience is considered as positive or negative determines the quality of this speech-related experience, and it is directed towards the talker. This book is about persons' evaluation of speech, may it be human natural or synthesized speech experienced in a passive listening scenario, or speech of a human or artificial dialog partner in an interactive conversation. It aims at identifying the most relevant factors that contribute to quality formation, and provides quantitative models to describe human quality ratings of talkers from speech samples and dialog partners in conversation. This perspective on the entity producing speech is reflected in the term *Talker Quality* in the title. This requires taking human cognitive processes into account, including social-cognitive associations and attributions.

The reason for addressing Talker Quality instead of, for example, signal or transmission quality lies in recent technological advances in signal processing and natural language understanding. This development has resulted in highly natural synthesized speech and high performance in automatic speech recognition. As a consequence, a new rise of speech interfaces is currently taking place, especially in (mobile) personal assistants and in smart-homes. Still, the user's perspective when evaluating speech interaction between humans and computers is not yet comprehensively understood. Especially with this emergence of new, voice-driven services, however, the interest increases to study the experienced quality of such voice-interfaces and in particular, how likable such a synthetic voice or the whole artificial conversational partner is perceived. Following the *Media Equation* [295], humans do treat media and interactive systems, especially voice user interfaces, as social actors—a notion that implies an evaluation of synthesized voices and voice-driven interfaces similar to interpersonal evaluations in humans. Although the equality of media and real life seems to be too strong to be supported unchallenged, several empirical findings presented in the referenced work and later on have proven that users do attribute human traits and states and intentional social behavior to interactive systems. Such attributions are based on surface features alone, such as the selected voice, although there is no social or personality component to explicitly model and exhibit dedicated behavior. In order to properly evaluate the user's perspective on such systems—especially with the recent increase in naturalness of speech synthesis and performance of such systems that become more and more similar to humans, this book approaches quality and evaluation from the perspective of human speech and human conversation: The idea is to identify processes and quality relevant factors in HHI, to transfer the results to HCI, and to validate them. This is also the reason for presenting studies in HHI here, as there have been open questions concerning the domain even in HHI.

This book focuses on first impressions and first encounters for two reasons: Firstly, studying familiar voices in humans may complicate things due to individual associations and expectations, and secondly, first impressions are very persistent [7, 23] and most relevant for future usage and thus also for acceptance of technology [32, 120].

After defining the most relevant terms with regard to *quality*, the theoretic groundwork for evaluating voices and voice-driven interaction is presented in this chapter. Most topics stem from the field of cognitive psychology. This includes definitions and explanations of impression formation processes and attributions of person states and traits (Sect. 1.2). The *Brunswikian lens model* will be introduced to explain human process of attribution. This model will be used in the successive chapters to associate own empirical contributions to identifying quality aspects and building quantitative models of Talker Quality to cognitive processes in order to define the status of each result. The development of first impressions over time is briefly outlined, along with the impact of overall evaluations, such as Talker Quality, on behavior. Chapter 1 closes with an introduction to anthropomorphic interfaces, which exhibit audio-visual speech synthesis.

The remaining chapters are more engineering-oriented by providing related work and own contributions on the evaluation of Talker Quality in passive listening scenarios (Chap. 2) and interactive conversations (Chap. 3). Here, empirical results and quantitative models are presented. The last chapter closes this book by interpreting the results from a development and design perspective, and by providing guidelines for application and future research.

1.1 Quality and Experience

Following established definitions from the research community dedicated to quality in telecommunication services, definitions related to the terms *perception, experience*, and *quality* will be presented.

> **Experiencing** is the individual stream of perceptions (of feelings, sensory percepts and concepts) that occurs in a particular situation of reference. [291, p. 13])

The authors describe *perception* on the same page as this definition as being "the conscious processing of sensory information the human subject is exposed to." Of course, perception is affected, even filtered, by attention [126]. The idea of perception being always conscious implies a separation of perception from pure sensation. For example, nonverbal signals (see Chap. 2) can stimulate reactions despite not being recalled later as perceived, and thus not being conscious [278].

As a consequence, these definitions limit the scope of experiencing to conscious or at least retrospectively assessable sensations. Nevertheless, the experience of a voice, a conversation, or the usage of a service does not comprise its evaluation. It is assumed that an individual appraisal of an experience is a separate cognitive process:

Quality of Experience is the degree of delight or annoyance of a person whose experiencing involves an application, service, or system. It results from the person's evaluation of the fulfillment of his or her expectations and needs with respect to the utility and/or enjoyment in the light of the person's context, personality and current state. [291, p. 19]

There are several important and noteworthy aspects in this definition of quality of experience (QoE). First of all, quality of experience is highly subjective, as the experience is evaluated with a very personal reference, comprised of expectations and situation-dependent needs. Therefore, the same stimulus may bear different QoEs for different people, different environments, or different times. Also, it is not explicitly stated whether QoE is conscious or not, but from the introduction of the term *perception* this can be concluded. It is also not stated, whether pragmatic and/or hedonic aspects are relevant (see below), but the general reference to need-fulfillment does imply that both might play a role according to the subject. In any case, QoE is directed to the "application, service, or system" [238, p. 5], and thus aims at optimizing the service in light of the users' perspective. Therefore, it is desired to exclude or control experiences from other sources that are not related to the service, for instance, the quality of the content provided by a media service, but not the media collection itself: For example, how entertaining a sports game or movie is, should not affect the QoE directed to the service, whereas the provision of a sports channel would be in scope. This kind of scope definition is not explicitly provided in the definition, though. And for interactive services or human encounters, separating form and content of the talker that is of interest here from other influences might not always be possible. For example, in the case of an electronic/human information service, a bad weather forecast should not result in a bad evaluation of the service.

That QoE is directed to the technology under evaluation is apparent in the scales used for explicit assessment: The studies conducted by the QoE community typically ask directly to rate the application, service, or system in question, instead of asking to report on users' general feelings [24, 181, 239, 315, 372]. The concept of quality of experience is also the target concept studied in this book, and its domain is the Talker Quality as used in the title. Therefore, a definition is provided that is adapted to the domain of spoken interaction and makes the relation to the talker more explicit.

Talker Quality is the degree of delight or annoyance of a person whose experiencing involves voice-based stimuli, either in a listening-only or in an interactive situation. Such stimuli may be produced by humans or machines, referred to both as "talkers". Talker Quality results from the person's evaluation of the fulfillment of his or her expectations and needs by the stimuli with respect to the utility and/or enjoyment in the light of the person's context, personality and current state.

When transferring QoE to human or anthropomorphic agents by adjusting its definition to Talker Quality, it resembles the concept of an interpersonal attitude a lot (Sect. 1.2), as it also addresses the experience-based evaluation of the talker, including individual expectations and presumptions. Also, attitudes "[...] are likes and dislikes" of a person, object, event, or place [29], in the case of this book of a person's synthesized or natural audio-(visual) speech. Following the definition by Eagly, ...

> [An attitude is] a psychological tendency that is expressed by evaluating a particular entity with some degree of favor or disfavor. [91, p. 1],

... an attitude is an experience-based latent (i.e., at first unconscious/implicit) evaluation that manifests in a specific situation [207]. Without going into details about the assumed cognitive architecture and processes involved in its manifestation, the notion of a "summary evaluation" concerning a given stimulus points out that it resembles a situated and individual quality evaluation, just like QoE. However, an attitude can also be a predisposition for a QoE evaluation, and its consequence. For the case of Talker Quality, a person may have, for example, formed an initial attitude, based on media information, towards a particular voiced-based smart-home device, which subsequently affects its Talker Quality in the first encounter. Similarly, the first experience with such a smart speaker will form a particular attitude towards this device. Conceptually, the term "attitude" stresses the latent character of an evaluation, while QoE aims at the situated "degree of delight and annoyance" directed to the talker; however, without neglecting predispositions and potential consequences. Already, research on QoE is expanded from single situations to multi-episodic evaluations [406], for which the latent character of attitude formation will increase in relevance. Coming back to the similarities between attitudes towards speakers and Talker Quality, liking is a well-fitting synonym for an attitude [106], especially the German term of "Sympathie" [149], which is even more important due to the cultural implementation of the own empirical contributions presented in this book. Liking and "Sympathie" are also well-fitting terms to express Talker Quality in its directed overall "degree of delight and annoyance."

Based on the previous statements and definitions of interpersonal attitudes from psychology, the following statement is derived to express the relationship between Talker Quality and attitudes for the purpose of studying and modeling the listeners' evaluations of voice-based stimuli in humans and machines from an engineering perspective:

> Attitude is a latent individual evaluation of another real or virtual person, developed on prior and/or immediate experiences and in relation to the person's expectations and needs. It can be expressed as summary evaluation on a positive-negative continuum in light of the person's specific context, personality, and current state. Talker Quality represents this situated expression of attitude.

With the definition of Talker Quality and by emphasizing its commonality with attitudes, the domain is explicitly tailored to the perception of human or human-like speech and to the spoken interaction with humans or human-like interactive systems. While QoE has already comprised interactive systems and services, interpersonal attributions, such as personality and attractiveness, but also stereotypical associations, are now taken into account during the formation process. The conceptual differentiation between implicit and explicit measures of QoE can now be applied by capturing conscious reports of Talker Quality, e.g., with questionnaires, and unconscious aspects, i.e., the attitude, by making use of reaction times, physiological measures, or behavioral observation (like amount of usage or spatial distance). An example of using distance measures is presented in Sect. 1.5. Interestingly, there are differences between implicitly assessed attitudes and explicitly reported evaluations, as implicitly assessed attitudes (e.g., with reaction time-based methods) have shown to be more relevant for actual and future behavior and are therefore more strongly related to acceptance of a service, product, or person. Typically, QoE is measured quantitatively and explicitly. Explicit ratings are also dominantly assessed in own contributions to Talker Quality. Exceptions from this are presented at the end of Sect. 2.1.

The separation of experiences into pragmatic and hedonic aspects of experiences stems from the evaluation of user interfaces. The term *user experience* is very similar to QoE and comprises affective, aesthetic, and social aspects (hedonic) as well as aspects of usefulness, efficiency, and effectiveness (pragmatic) [144]. The difference between user experience (UX) and QoE lies in the focus, i.e., its application: Whereas QoE is used to evaluate (telecommunication) systems and services from an engineering perspective, and thus emphasizes overall (quantitative) ratings of good–bad of the service implementation and interface, UX was established especially for the domain of user interface research/design. One of its aims is to optimize the user experience by improving the interface, at least more importantly than optimizing the service implementation [216]. Of course, with the recent developments in, e.g., smartphone applications and chat bots, developing innovative small-scale services that fulfill everyday needs are entangled strongly with the interface concept. In contrast to QoE, the development of UX over the time of exposure and interaction represents an important research and evaluation topic [208], and qualitative methods are often applied to study and assess UX [57, 145, 380]. However, studying the complex nature of experiences and the individual aspects apart from an overall good–bad dimension is of interest for both communities.

Like many other concepts with practical implications, UX is not consistently defined and assessed [215]. However, one established definition reflects common ground—maybe because it lacks details—and illustrates the holistic perspective of UX:

> **User Experience:** A person's perception and responses that results from the use or anticipated use of a product, system or service. [157, p. 7]

A central term in this definition is the usage itself, may this being in a passive or active scenario [303], whereas QoE focuses on the resulting evaluation of such usage. Also, user experience includes perception, i.e., also unconscious aspects that may affect emotion, mood, etc. Despite UX apparently being the more "mature" concept [376], these differences in scope and focus are favoring QoE as central research concept in line with the aim and focus of this book.

Kahneman distinguishes between a momentary-based approach and a memory-based approach in human experience [177]. The momentary-based approach reflects instantaneous experiences that can be attempted to be assessed immediately. Its variation over time corresponds to macroscopic changes of, for example, system characteristics and subsequent events in interaction. Accumulating ratings of momentary experience, for example, by averaging, represents useful abstraction, not actual experience, called total utility. Assessment methods of momentary QoE include those for truly instantaneous assessment, but also "sampled momentary" QoE, i.e., assessment for short-term samples which exhibit no macroscopic, but microscopic variation. The memory-based approach is concerned with retrospective appraisal, i.e., individually remembered experience. This remembered utility reflects the users' integration processes, including memory-based biases and filters, which affect the establishment and development of attitudes towards a system or service. The majority of studies presented here, including the ones conducted by the author, aim for the memory-based evaluation of talkers that is assessed retrospective, explicit ratings directly after experienced encounters. Other aspects, such as momentary experiences that are a relevant aspect of user experience, are not in scope, as they are not defined contribute to the overall evaluation and Talker Quality. Still, why is QoE a relevant topic for engineering, i.e., designing and developing interactive systems that allow for spoken conversations? In contrast to QoE, there is already a long-established engineering concept, the so-called quality of service (QoS):

Quality of Service: Totality of characteristics of a telecommunications service that bear on its ability to satisfy stated and implied needs of the user of the service. [161, p. 2]

The perspective is, however, concentrated on the system developer or service provider. Even users' needs are addressed in a general and technology driven perspective, as they can be only implied. Therefore, the scope of this definition is the technical performance of a service or system in order to optimize it [216], as precondition to QoE ("its *ability* to satisfy") (italics inserted), and thus excludes those individual aspects of QoE dealing with individual and subjective needs, for example, aesthetics and style, identification, or affect. Therefore, QoS and QoE establish a meaningful duality, with QoS directed to optimizing the performance, and QoE the user perspective, and ultimately, the user acceptance. Since QoE is addressing the situated and individual aspects as the definition of "experiencing" states, human social cognition (Sect. 1.2), for example, effects of social actorship in HCI (Sect. 1.4) have to be assigned to the research agenda of Talker Quality that is dealing with human and human-like speech.

1.2 First Impressions and Attitude Formation in Humans

Attribution processes deal with inferences about people that are based on limited information and prior experience (e.g., schemata or stereotypes), applying heuristics to come to conclusions quickly with low cognitive load. Such attribution processes are especially relevant and apparent in first encounters, due to the necessity for appraisal of limited information to practically deal with each other (cf. [15] for an overview). Therefore, these first impressions do affect how relationships might evolve in the future [171]. Attribution theories deal with the way people explain observations (nonverbal and verbal), thus estimating and guessing traits and states of persons. These explanations can be separated into internal attributions (the origin of the observed behavior is assigned to the speaker, e.g., the personality) and external attributions (observed behavior is assigned to other factors, e.g., the situation or social circumstances). For first encounters, internal attributions are rather frequent.

speaker traits: comprise slowly changing characteristics, such as age, gender, social group and regional background, speech pathologies, personality, inferred sexual attractiveness

speaker states: include passing characteristics, such as mood and emotion, intoxication, even interpersonal aspects (friendliness, interest)

For the development of relationships, there are different phases postulated [221], beginning with first *acquaintance*, *buildup* of trust and intimacy, *continuation* towards a strong and stable relationship, and a potential decline with a *deterioration* and finally the *termination*. Originally modeling romantic relationships of couples, this and related views also work for friendships of two or more people [194]. Common to such views on relationship development is the strong impact of the first impression to continue to later stages. For potential friendships or even business relations, a positive first impression is mandatory. The first phase of *acquaintance* (or *initiation* [192]) is characterized by a first interaction or even one-sided perception. It is based on (mutual) attraction or interest. For the scope of this book, this first phase is most relevant, as it represents both, the passive and the initial interactive zero-acquaintance situation, when no prior experience is available.

In the light of relationship development, the first encounter is mostly about the final evaluation of the first impression, whether the other is socially attractive [154]. Social attractiveness can be considered as a major interpersonal evaluation that is based, within the scope of this book, on speech experiences. Whereas attributions can be positive (e.g., intelligent), negative (e.g., lazy), or even neutral (e.g., age group), liking somebody or not is a subjective assessment. In contrast to attributions that estimate persons' states or traits, this social attractiveness can also be treated as an attitude of the evaluator towards the other person [149], and is consequently also termed *attitudinal positivity* [154]. For a real persons' voice, social attractiveness is therefore identical to Talker Quality and in the referred studies on attitudes towards human speakers (Sect. 2.3), the term "social attractiveness" is quite present. In addition to the engineering perspective outlined in Sect. 1.1 that is expressed in the definition of Talker Quality, however, favoring Talker Quality over "social attractiveness" in the Chaps. 2, 3, and 4 on vocal aspects also avoids a misinterpretation of the latter term as social prestige that is an important factor for technology acceptance (Sect. 1.5): While with prestige "social" refers to the peer group or society, social attractiveness addresses the social aspect of the pair of people. Traditionally, an attitude consists of the three aspects,

1. affective reactions towards the stimulus,
2. cognitive associations, such as beliefs about the stimulus, and
3. behavioral reaction, e.g., your verbal and nonverbal interaction with that person [106, 207].

However, the current perspective is generally concerned with a tendency to evaluate the stimulus on a valence dimension, i.e., a summary evaluation that could also be called "preference," and it is, like any overall evaluation, individual and self-related [106]. For the case of the first vocal encounter, this attitude and thus Talker Quality is still based on very little information. For participants in experiments, there is therefore reason to expect a high degree of rating consistency between people with similar background. How this consistency diversifies over time is presented for social attractiveness in Sect. 1.3. Nevertheless, Talker Quality is an interpersonal attitude, and therefore not only depends on the stimulus characteristics, but is also subject to the rater's attributions and expectations. Therefore, it can be assumed to comprise common, but also individual components, e.g., references and associations that affect the assessment of another's voice. One quantitative approach to study individual components is presented in Sect. 2.1.

Attitudes can be assessed explicitly (i.e., directly), e.g., on ratings scales or preference choices, qualitatively by methods like the Repertory Grid Technique [183] that elicits individual labels, or truly implicitly, e.g., by the Implicit Association Test [207] that collects reaction times. Other approaches apply physiological or even neuro-physiological measures. Dual-process models deal with implicit and explicit attitude formation: Common to different dual-process models is a separation between the direct effect of peripheral or superficial information, such as aesthetics or other traits (of a person) on the attitude towards a given stimulus, and the effect of reasoning and consideration of the central information at hand, e.g., especially taking into account the most relevant information for the evaluation in question [106, 274, 281]. An example would be the attitude towards a product being based on the persons' appearance in a commercial (e.g., sexual attractiveness, regional background, or ethnic group), also called peripheral route [281], compared to the functionality and price of the product. The peripheral route is considered as cognitively economic, fast, and automatic, and can build on schemata like stereotypes. This implicit attribution process may be automatic. However, it can be changed by an additional conscious and explicit process, if we (can) make the effort of reflecting on the information available or even gathering missing information [123]. This explicit process is also called the central route [281], which may be more costly, but it can be activated with sufficient motivation, cognitive resources, and ability.

Especially when based on little information, however, first impression of unacquainted persons will be "quick" attributions and will be dominated by few salient features that allow assigning the current person/voice to already existing categories to adopt the attitude from there. As this first step is regarded as automatic and implicit, it seems unavoidable per se. But it is not determined, how strong this implicit attitude is and how easy this first impression can be revised. Indeed, this kind of first impression has been found to be quite persistent and not easy to overcome or influence even with contradicting information [139, 341], with all its potential to become a self-fulfilling prophecy by one's own actions.

If the second, conscious, processing step is not taken, then so-called fundamental attribution errors can occur: Due to over-generalizing, internal attributions might be preferred over (more realistic) external ones. One example would be the attribution of personality caused by the regional background that is audible in pronunciation and prosody, and could be completed with a subsequent positive or negative assessment. Attributions that are invalid, but still consistent, can be explained by applying stereotypes and attitudes towards categories (like regional heritage), which are both socially grounded and can therefore result in reliable judgments.

Common domains to study dual processes are advertisement and persuasion, and in particular social evaluations of competency, intelligence, and of course, social attractiveness [112]. The approach of dual processes is presented here for two reasons: Firstly, it emphasizes the relevance of superficial information, such as from the voice and face, on Talker Quality for first encounters. Dual-process and related models can explain why evaluations of speech are governed by state and trait attributions towards speakers, and why aesthetics play an important role in global evaluation. Secondly, it explains the potential discrepancy between implicit measures of attitudes and explicit measures of attitudes, as the implicit ones resemble the peripheral, affective process, whereas the explicit ones could be adjusted to social norms when reporting.

The attribution process in humans can be illustrated by Brunswikian lens model (Fig. 1.1, here in its modification by Scherer, separating two stages/levels

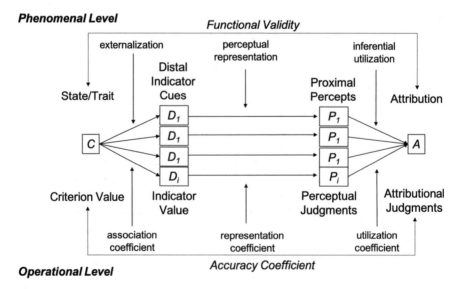

Fig. 1.1 Scheme of Scherer's modified Brunswikian lens model, adapted from [310]

of cues [310]). The basic idea is the distinction between speakers traits/states that are externalized into (some) *distal* cues and perceived by other as *proximal* percepts. Externalizations can be directly measured by physiological, articulatory, or simply acoustic or visual means; in the scope of this book acoustic parameters of paralinguistic information and interaction behavior are taken into account. Externalizations of states/traits may be perceived by other persons, e.g., interaction partners, and subsequently used for attributing states/traits to the speaker. These *proximal* percepts represent the perception of such physical cues in form of perceptual categories or dimensions. The level of human percepts is reflecting filtering and selection processes due to human physiology and cognitive processes. This level of human perception is usually explicitly assessed by asking participants in experiments. For the case of human voices the perceptual representation might be, for instance, the perceptual dimensions of perceived pitch and vocal effort. Obviously, this model describes a probabilistic process, as the state/trait cannot be directly observed, but has to be inferred by the information available. Which cues are taken into account as evidence is moderated by perceivers' expectations, experience, and perceptual salience. Therefore, the lens metaphor uses the *focus* to illustrate the selection and weighting process.

Voice experts, for example, might have different perceptual dimensions than ordinary people, just due to their training and knowledge of articulation and physiology, and the decision on the methodology and participants should be taken with the research question in mind, e.g., studying the externalizations by favoring experts vs. studying perceptual dimensions and attributions by favoring normal participants.

The fact that persons do strongly and actively attribute states/traits, especially in the case of sparse data in early and brief encounters, is the origin of specialized studies on perceptual dimensions of voices. Results on perceptual dimensions for speakers are presented in Sect. 2.2.

According to the model, from these perceptual dimensions a selection is evaluated to infer a trait or state: this is the actual attribution. This selection and evaluation process might result in invalid attributions, e.g., by applying a stereotype of males being more competent than females, thus taking into account an invalid cue for the target trait (in this example the gender for estimating competence). On the phenomenal level, if the cues for externalization of a concept match the ones used by others during perception, then this concept is validly attributed by other people for a given medium (upper path of the figure). On the operational level, the degree of validity is reflected by the chosen selection and representation of the cues as coefficients (lower path in the figure). Typically, quantitative modeling of attributions from perceptual representation is done in a linear fashion by applying weights. However, other, non-linear relationships might be also appropriate, e.g., in cases of one percept masking another due to its intensity/salience.

Scherer has applied this lens model to the domains of emotion and personality in vocal communication. But others, especially physical traits, have also been studied. Some of these concepts in question (the state or traits denoted as C in the figure) can be easily measured: For example, body size or chronological age provides a good ground truth to be examined. Others, like personality or emotion, are more difficult. For personality, for example, close friends and relatives can be asked to average questionnaire data, which is not possible for self-reports. Emotions might be stimulated with experiments, but with limited certainty about the precision of the state information.

Although this model covers the whole interpersonal attribution process, also parts of it can be applied to meet specific scientific goals, such as synthesizing cues with the aim of stimulating attributions—with a focus on synthesizing the relevant percepts, or recognizing human states/traits by concentrating on the externalizations. For the latter, the attribution A would be the ground truth to meet. Therefore, it is of interest to separate studies on validity from studies of consistency and reliability:

- Validity or accuracy is concerned with the relationship between the state/trait C and the attribution A.
- Consistency between raters of A and reliability (result similarity for repeated or similar tests) informs about the status of the attribution concepts—whether it can be considered as manifest or not.

If a trait/state has low validity but high consistency, for example, it would be a stereotype. For vocal cues, physical traits like age group, sex, physical (visual) attractiveness, body size, and body weight are consistently attributed [203, 356, 417]. However, the validity for body size is rather low. For concepts other than physical ones, the validity of vocal cues is typically much lower and supports only selected person information, e.g., the extroversion dimension of personality [310] or the arousal dimension of affect (cf. [128]).

Recent work on automatically estimating concepts of self-revelation based on acoustic features includes age group, sex, body size, physical attractiveness, personality, and even native language with quite success. The domains include health topics [e.g., depression, autism, Parkinson's, and forensics (e.g., intoxication)] [317]. In the series of the Interspeech Computational Paralinguistic Challenge, these externalizations are mixed with listeners' ratings of, e.g., personality or likability.

What is actually missing in the figure for modeling Talker Quality is a final assessment of the attributions made. Such an extension of the model should consider Talker Quality in terms of liking vs. aversion, but might also include the consequences of the inference process, such as affective and behavioral (re)actions (e.g., willingness and probability of initiating future interactions). The lens model builds the methodological basis of this book insofar as following it allows for developing informed quantitative models of user ratings by incorporating existing knowledge of relevant measures for externalizations, perceptual dimensions, and interpersonal attributions, as well as empirical methods to identify missing parameters and concepts required for the models.

Apart from the already mentioned speaker attributions of traits and states, there seems to be two basic and universal attributions for (human) interpersonal perception, even for short stimuli. Although named with many different terms, the first dimension is related to *interpersonal warmth and agreeableness* or *communion*, whereas the second one is related to persons' capabilities, competence, and *agency* [1, 113, 307]. For the domain of speech stimuli, these dimensions are often called benevolence and competence.

Despite this resemblance of the first dimension (warmth, benevolence, etc.) with social attractiveness [13, 86, 113], liking–aversion may conceptually also comprise physical attractiveness (or its inference from vocal cues for Talker Quality) and competence [246], even in speech [289]. Actually, there is much evidence from questionnaire analysis during dimension reduction that evaluative questionnaire items can be apparent in both dimension, warmth and competence, or neither [47, 48, 74, 142, 337, 382, 386]. Therefore, it remains unclear whether the claim of the warmth dimension of interpersonal perception being identical to social attractiveness is true. Given the empirical evidence, it can be argued that the so-called benevolence is just one possible and very likely attribution affecting the attitude formation and Talker Quality, especially in a first impression. Talker Quality is defined in the frame of this book, not a feature of another person, such as the attribution of benevolence or competence, but instead an individual attitude towards this talker. This is in line with the correspondence of attitudes with valence of a three-dimensional approach on connotations of meaning, comprising valence, dominance, and activity [74, 270] that is also applied to define emotions [16, 253].

1.3 Attitudes: Long-Term Evidence and Validity

From already established human relationships, four relevant factors are known to affect social attraction [15]:

Physical Attractiveness: Beauty is a typical cause for halo effects, i.e., a judgment or attribution based on an unrelated cue, resulting in a general positive attitude towards that person and increased likelihood for a lasting relationship.

Similarity: People with similarities in subculture (e.g., visually expressed like clothing), hobbies, background, or personalities are more positively evaluated.

The Propinquity Effect: Various kinds of familiarity, and may that be mere exposure to that person, can result in a higher social attraction.

Reciprocity: We tend to like people who display to like or be interested in us.

The question is, to which amount these systematized findings are relevant for Talker Quality? Physical attractiveness estimated from acoustic, vocal cues does represent a consistent, even a valid first impression [69, 70, 102, 417, 418]. This feature will therefore play a role for Talker Quality of unacquainted voices and will be reflected in acoustic features. Similarity in regional and social background can also be quickly assessed and analyzed to explain or control for individual sources of variance (refer to Sect. 2.2), but requires some effort in recruiting. Much more interesting from a speech signal processing point of view is the case of acoustic similarity, which failed to show an effect, and will be presented along personality similarities in Sect. 2.1.

From a methodological perspective, there should be no Propinquity Effect active for first impressions, as voices should be presented either only once or with the same number of occurrences/repetitions in passive rating scenarios. Still, a voice reminding to a well-known and positively regarded person might give rise to such a familiarity, but this kind of artifact should be routed out by averaging. The case of similarities to famous voices like actors might of course be a methodological issue, but up to now, personal experience in obligatory post-experimental interviews did not include such an effect.

In contrast to propinquity, reciprocity is definitely relevant for unacquainted voices. Hearing somebody smile, or other signals of interest and benevolence, even an open or extrovert personality, might all be perceived as positive attitude and subsequently trigger reciprocity, even in a passive experimental scenario. One example might be the case of the Interspeech challenge speaker likability database, in which users operated a telephone voice portal [318]. Caused by the task, many speakers used a command style in this particular situation. However, despite this style being a frequent and normal behavior, it was negatively evaluated compared to a more conversational style, which was interpreted by the authors in the light of reciprocity [388]. In first encounters, there might be sufficient information missing in order to validly explain vocal signals. Consequently, there could occur attributions errors in the presence of a specific situation (in the voice portal example, there might have been a negative attitude perceived).

Whereas personality is not explicitly mentioned to play a crucial role in the four factors of long-term attraction, there are results indicating a general preference of specific personality profiles, i.e., extroverted and agreeable persons, who are being liked with higher probability (cf. [355, 412]). This result might be caused by a positive impression (displaying benevolent or friendly behavior) seemingly directed to the (experimental) raters, who would be affected by reciprocity.

Reciprocity is apparently highly affected by degree of acquaintance. With the *Social Relations Model* [186], interpersonal ratings within groups can be used to statistically separate the three sources of variances:

1. The target variance (how the person rated is usually liked on average).
2. The perceiver variance (how does the rater usually rate on average).
3. The interaction variance (what is the unique effect of the individual dyad in addition to the target and rater effect).

For social attractiveness (assessed with the term "liking"), about 10–20% of the data variance can be attributed to the target and the rater each, whereas about 40% are specific to the dyad [186]. The latter exemplifies the relevance of rater's individual perception and attribution concerning the specific target (speaker) and the low consensus for such ratings of acquainted people. With data analyzed by this model, reciprocity of liking ratings (or affect, as it is called in [186]) increases with acquaintance: The better, or longer, people know each other, the stronger is the correlation between the interaction effect of A towards B and B towards A. In contrast to traits of interpersonal perception, such as personality attribution, this effect increases from rather low to high reciprocity (e.g., from $r = 0.26$ to 0.61 for the data presented in [186]).

The relevance of the first impression is not only evident in relationships, but also in terms of validity of performance prediction. For HHI, a research community has been formed that studies the validity of the first passively obtained impression from so-called "thin slices" for longer-term evaluations. Thin slices are all kind of stimuli that exhibit a person's dynamic, especially nonverbal behavior (e.g., content-masked prosodic information or body movement, but not still pictures) and shorter in duration than 5 min [8]. Following this definition, all spoken stimuli can be considered as thin slices due to their inherent nonverbal content, perhaps excluding only steady-state vowels or heavily edited speech material. It could be shown that acoustic samples of 20 s already provide sufficient information to form a judgment of salesmen's skills and anxiety: "Thus, thin slices force the observer to focus on nonverbal cues without the influence of the verbal message or information from previous interactions of the broader context of the situation." [8, p. 5]

The implications seem drastic: Is nonverbal—in our case of speech this is mainly prosody—information interpreted consistently by various listeners without content and situation? Is this information truly predicting (or representing) future performance and evaluations based on sufficient data? Several studies show that 20 s is a proper time span to predict job performance and judge aspects of skills and personality reliably and accurately, although in most studies audio-visual information was used, not only audio—always assuming representative slices (e.g., not highly emotional parts, when the context is basically factual) [5]. Even shorter slices down to 5 s are used to obtain reliable results.

The validity of many studies of such impressions and judgments has been collected and analyzed [5]. Despite being dependent on various factors, different variables/concepts can be assessed with thin acoustic slices. For the case of the scale "likability," reliability remained rather low both for the auditory and for visual channel (Spearman–Brown reliability of $\rho = 0.188$ and $\rho = 0.171$, respectively). Correlations between ratings of thin slices and performance indicators after longer time periods are quite high for some domains, e.g., school teachers [6]. Still, the origin of this relationship, may it be a persistent first impression (attributions) or actual validity of representative parameters, is to be discussed for such scenarios. Clearer and even more impressive results have been obtained for thin slices used to predict partnership status years after the thin slices' evaluation [127].

Other examples regard structured interviews with managers [82], recordings of salesmen [280], telephone-based costumer support [146], and employment negotiations [76].

Apparently, there are not only several attributions, but also evaluations triggered by short vocal stimuli that show consistency over raters in this passive scenario, and even show persistence over time and validity concerning performance measures. The persistence and validity of the first impression is even studied for HCI (Sect. 3.4.2).

1.4   Quality of Experiencing Anthropomorphic Interfaces[1]

Anthropomorphic user interfaces are human-like computer agents. Due to their human-like usage of spoken language, voice user interface could also be counted among anthropomorphic interfaces, but typically, this term is applied for the visual domain and comprises humanoid robots and virtual humans that are presented on displays as two- or three-dimensional models. For virtual humans, the visual models might represent a complete body, or just the head or torso when (audio-visual) speech interaction is in focus. Such virtual humans that have the capability of spoken (and nonverbal) interaction are also referred to as embodied conversational agents (ECAs).

The formation process of a first impression outlined before for speech-only experiences is in principle not different to that of audio-visual speech stimuli. The perception process, of course, can be much quicker for purely visual compared to auditory information, just because a static visual stimulus is already two-dimensional and is coded in a much higher frequency domain than acoustics, and thus might hold sufficient information for inference processes. In order to elicit such automatic attribution processes, researchers even explicitly limit the exposure time of pictures of websites to tenth of milliseconds (see Sect. 3.4.2) or to 100 ms for faces [350], whereas for voice stimuli, typical segments are vowels and syllables with minimal duration of about 150 ms, but mostly longer, up to several seconds for sentences [105]. As virtual characters and social robots are designed to independently engage in social and verbal interaction, they are usually called agents. Talker Quality in interactive scenarios is covered in Chap. 3.

With the ongoing development of anthropomorphic agents, the need for appropriate evaluation methods for such user interfaces increases. Exploring the field of synthesizing required personality traits and affective states by appearance, voice, and communicative behavior is of recent interest. Basic decisions concern the fit between visual and auditory information, as well as between the attributed states/traits and the expected role of such an interface.

More elaborate issues of Talker Quality are concerned with the social aspects of the first impression. That users may treat such agents as social actors represents one major reason for choosing anthropomorphism in the first place. One common term covering such behavior is "believability," and does not imply that an agent might be indistinguishable from a real human. It only means that users show behavior as when dealing with humans, e.g., increase number of social signals that is caused by anthropomorphism, and which seems to be an automatic response to an artificial social actor [329, 361]. Of course, such behavior can also be triggered by other causes, such as non-anthropomorphic new media [295]. But typical system features relevant to increase believability are nonverbal signals, especially gaze, facial expressions, and communicative gestures, but also politeness expressed verbally

---

[1]Parts of this section have been published previously [402].

[201, 363]. Believability, warmth, and competence, assessed with rating scales, for instance, were improved by plausible and appropriate, and multimodal emotion display of an ECA [85].

Consequently, user tests for this aspect are a necessary, but not sufficient condition for a positive Talker Quality, at least in relation to the anthropomorphism. Effects known from social psychology can also be utilized to assess this aspect of quality in passive and interactive scenarios [402]. Already for the passive case, automatic person attribution processes, concerning, e.g., intelligence, competence, or stereotypes, would indicate that an agent is a believable social actor and thus treated as a potential conversational partner. For the interactive scenario, there are four effects especially relevant for HCI, with social facilitation/inhibition and the equilibrium effect indicating social actorship in general, whereas the latter two effects already indicate a level of quality of the user experience:

Social Facilitation/Inhibition: Two complementary observations have to be made in order to eventuate this effect: Performance of a person is enhanced in the presence of other social actors for tasks of low complexity, while performance decreases for difficult, complex tasks. One explanation of this effect is an increase in attention and arousal due to the social situation [135]. Other social actors instead of people can be anthropomorphic interfaces or social robots in the case of HCI. So, if the social facilitation/inhibition effect is found for a specific interface or robot, this would indicate that the particular agent is treated as social actor [300, 329, 377]. Social facilitation might be more likely for ECAs with gender according to the sexual orientation of the user [179] and be affected by social signals, as user-directed gaze resulted in social facilitation/inhibition with a humanoid robot [330].

Equilibrium Effect: Actually called equilibrium theory, this effect describes human social behavior to balance social signals of intimacy, i.e., social distance, on various modalities, most popular for proximity and gaze [14]. For HHI, the elevator exemplifies this effect, where the environment does not allow an appropriate spatial distance. As a reaction, an appropriate level of intimacy is restored by averting eye-contact. This assumption of balancing social signals has been tested especially for virtual environments. Empirical results show effects of equilibrium and a general tendency to manifest a personal space in contrast to non-anthropomorphic interfaces for ECAs and robots [19, 196, 262]. This also means that a comfortable distance for users is mediated not only by physical distance, but also by other indicators of intimacy, such as gaze [343]. Using a psychophysical distance measure instead of a purely spatial one improves modeling spatial behavior between humans and robots [251]. However, this multimodal compensation mechanism seems to be also moderated by likability, as it does not occur with benevolently introduced robots [257].

Uncanny Valley: This effect describes an increase of the perceived familiarity with increasing resemblance to human appearance only until a certain level of human-likeness is reached. However, additional improvements towards only a small divergence from human-likeness results in an uneasy feeling. At that level, familiarity drops and increases again only if the human-likeness is perfect, or at least very good [248]. The dependency of this effect on the expectation of naturalness is assumed to be caused by a change of reference from a technological interface to that of a real human. This effect provides a delicate challenge for the developer/designer to find the optimal configuration for anthropomorphism and might be countered by switching to non-natural (e.g., comic-strip like) representations.

Persona Effect: A positive effect of choosing a human-like interface metaphor compared to a non-anthropomorphic metaphor with the comparable service functionalities, e.g., a GUI, is observed. But even an accompanying ECA, e.g., to a website, might have a positive effect on users. It is mostly observed in explicit ratings, but also in increased measures of efficiency and effectiveness. However, many studies show that this positive effect is not inevitable. In fact, it is highly debated, as it seems to be dependent on task, system, and situation [83, 116, 416]. Nevertheless, from a QoE point of view, the appearance of the persona effect is the ultimate test of the design decision in favor of anthropomorphism. The rational is that users can apply (parts) of their livelong learned conversation competences in HHI to this anthropomorphic agent and do not need to know specific domains or interfaces [344]. However, it is still unclear whether the persona effect is related to social facilitation or indeed a consequence of a more natural and intuitive interaction. At least, compared to pure spoken dialog systems, ECAs can facilitate interaction [88], as human processing benefits from multimodal signals in decreased load and higher neural activity (super-additivity) [271, 331]. So, if not engaged in multiple tasks, an ECA might be less demanding to interact with, if the nonverbal signals are produced and recognized well. Despite this assumed and also observed persona effect, the design decisions and evaluation concepts concerning ECAs are not to be taken easily: Natural interaction can only have its positive effect, if it is supported well by the ECA, and visually human-like ECAs may fuel too high expectations for the dialogic and nonverbal capabilities. Although a natural interaction is aiming at establishing a social situation, including users do display signs of social phenomena, some users may like to avoid such social interaction for certain tasks in favor of a paradigm of direct manipulation; just like some customers prefer anonymous online shopping or self-service to personal service. As anecdotal benefit of an audio-visual speech interface over a non-embodied smart environment, users immediately have a clear direction to direct their speech to, resulting in a higher speech recognition rate.

To make use of reciprocity, anthropomorphic interfaces should express a positive attitude towards the users. Synthesizing such an attitude is typically based on nonverbal information transferred from HHI. For example, the impression of a conversation partner signaling interest, liking, and intimacy is correlated to smiling, eye-contact, forward leaning posture, and proximity [53, 115]. However, implementing likable and dislikable conditions can also be aimed by means of benevolent vs. impolite verbal behavior, e.g., in a robot experiment [257]. Modeling positive vs. negative attitudes in a virtual agent by nonverbal and verbal signals was, however, only successful for the nonverbal information [64].

Which kind of ECA or robot is preferred and liked is, however, still an open topic, as it is for voice synthesis alone. Smooth compared to mechanic movements increase likability for robots [65]. Other results indicate that both synthesized attitude and personality affect social attractiveness. In a series of three studies on first impressions, for example, a synthesized friendly attitude towards human museum visitors was stronger stimulated by smiling and eye-contact than an extrovert personality that was expressed by virtually stepping forward a single step on the screen's virtual space [62]. This friendly attitude, but not the extrovert personality, was found to be closely related to the virtual museum guide rated as approachable and likable. Consequently, for the friendly ECA version, user reported a higher preference and a higher willingness of future interactions. In the final field study in a real museum, the friendly version attracted slightly more visitors, but this difference was not significant.

Up to now, virtual persons have not been distinguished from robots in this section. However, it has to be mentioned that robots do have drastic technical limits when it comes to appearance and visible movements. Countering these limitations, robots exhibit a physical presence, i.e., the possibility of real interaction with persons and the environment. This physical manifestation also increases social presence [184, 219]. Results of a meta-analysis show a better user performance and more positive attitude towards co-present robots compared to virtual agents and tele-present robots [226].

1.5 Quality Affects Behavior

Quality evaluation of speech is not an end in itself. When creating synthetic speech or developing a conversational system, it should be used. Here comes the behavioral component of an attitude into play (Sect. 1.2): The evaluation of a first impression affects actual behavior towards the interaction partner [304], not only in terms of reciprocity of social signals, but also concerning the probability and frequency of following interactions [149, 204, 301] and thus the formation and development of relationships [221, 341]. In HCI, for example, the first passive and interactive impression might already establish a barrier or carrier for future usage [120].

For the relationship between initial evaluation and interactive behavior in HHI, there have been four models proposed for dyadic social distance signals, i.e., how much distance interaction partners establish between each other or by number or duration of gaze: (1) reciprocity, (2) compensation, (3) attraction-mediation, and (4) attraction-transformation [178]. In terms of distance behavior, such as physical distance, but also gaze signals, or verbal disclosure, the four models would predict differences in relation to another's distance, sometimes moderated by the attitude "liking": Reciprocity and compensation are both independent from likability and just model the distance signals (approach or withdrawal) according to the interaction partner, either by mimicking the other (reciprocity) or by balancing the other to preserve a stable sum of distance signaled (compensation). In contrast to these two, attraction-mediation and attraction-transformation are affected by likability. Attraction-mediation describes self-distance signals according to likability independent of the other's behavior—likable causes approach, while unlikable leads to withdrawal. The last model, attraction-transformation, assumes reciprocity for likable partners, but compensation for unlikable ones. Empirically testing these four models in comparison, verbal disclosure followed a reciprocal pattern regardless of the liking, whereas nonverbal gaze was dependent of liking: for a liked experimenter, intimacy signaled with gaze was reciprocal, but balanced for a disliked conversation partner, which follows the attraction-transformation theory [178]. Such studies have also been conducted for human–robot interaction [257].

From a methodological point of view, passive scenarios cannot capture behavioral changes, which might limit the validity of explicit ratings, when it comes to real-usage Talker Quality, and ultimately user acceptance. In interaction experiments, individual differences might already induce variance for the first encounter. In order to cope with such effects, especially when aiming for quantitative models, inter-individual differences should be taken into account, for example, by applying non-linearities as covered by the attraction-transformation model.

For spoken and face-to-face interaction, this behavioral component of attitudes has severe implications. For example, user differences are often not taken into account (exceptional examples are [199, 364, 381, 397, 400]), as the perspective of evaluating a system seems to hinder considering user and computer as coupled in a relationship. For example, parameters describing the course of an individual interaction for predicting Talker Quality do not only reflect the performance of a system or service, but also the user's attitude and expectations.

So far, technology acceptance research has aimed at identifying relevant factors to increase or hinder the usage of new technology; especially for professional users. The most popular theories apply questionnaires and work on rather abstract levels, e.g., taking into account the perceived usability of the program or device in question (neglecting here ethnographic approaches that do conduct case studies). In contrast to this subject area, QoE researchers operate on a lower level, e.g., by studying real interactions to explain the emergence of individual and aggregated evaluations and attitudes: Estimates of user acceptance are rather build bottom-up, just like assessments of QoE or usability issues are directly linked to concrete incidents of experience. Therefore, technology acceptance models might be helpful to validate the design and procedure of empirical interaction studies.

From a series of studies [79, 209, 358, 359], the following factors potentially affect user acceptance or at least future usage avowals: perceived utility, perceived usability, perceived value for money, perceived appearance, social importance (prestige), voluntary nature of usage, identification with the service/product, perceived control of the service/device, emotional reaction to its use, and stimulation in use. The different models proposed base on the theory of reasoned action [3] and make use of structural equation modeling of questionnaire data. For the purpose of modeling Talker Quality in verbal interactions, monetary aspects are typically discarded. Still, established aspects of HCI-related evaluations have been identified as relevant and included into the models. These aspects are, for example, represented in the questionnaire AttrakDiff [144], which is often applied to evaluate HCI. This questionnaire for overall attractiveness covers factors of usability/utility, stimulation (emotional reaction to its use and stimulation in use), and identification (perceived appearance and identification with the product).

How Talker Quality affects conversational behavior of persons plays only a minor role in the empirical work, and own contributions on this topic will be presented in Sect. 3.2. Before that, methods, state of the art and own contributions on perceptual dimensions, and Talker Quality are studied in passive scenarios (Chap. 2).

Chapter 2
Talker Quality in Passive Scenarios

In passive scenarios, humans encounter a (virtual) person for the first time and have to create an impression based on sparse surface information. Experimentally, such a scenario can easily be controlled by inviting people into a laboratory and asking them to listen to a recording of unacquainted (virtual) people. This experimental research paradigm of passive scenarios is often applied in order to systematically pre-select the encounters, to adopt specific assessment methods, and to control for environmental effects, such as social situation or environmental noise. Most common is to ask participants for their explicit first impression with a questionnaire or a single ratings scale. Validations of such measures as well as further evaluation methods will be presented in Sect. 2.1. This is succeeded by the empirical work on perceptual dimensions, and on Talker Quality directly.

2.1 Methods and Instruments for Passive Scenarios

2.1.1 State of the Art in Methods and Instruments for Passive Scenarios

Before presenting own contributions to the identification of acoustic correlates of social attractiveness, the issue of the psychometric properties of reliability and consistency of the typical instruments applied has to be dealt with. In the studies referenced so far, social attractiveness has been assessed with a single rating scale (likable–not likable) or on multiple items of a full-fledged questionnaire. As this person-related subjective evaluation is concerned with an overall quality, a single scale should be sufficient to collect the liking or aversion towards a speaker, so that one or two additional items might just be used in case of improving validity and reducing noise that might be introduced by the labels [186, 320]. For identifying perceptual dimensions, as used in the lens model approach (Sect. 1.2), rating stimuli

© Springer Nature Switzerland AG 2020

B. Weiss, *Talker Quality in Human and Machine Interaction*, T-Labs Series
in Telecommunication Services, https://doi.org/10.1007/978-3-030-22769-2_2

successively on a single scale is of course not meaningful. For that aim, full questionnaires or other evaluation methods are required and the application of such instruments is presented in Sect. 2.2.

Especially for the German case, there is a fitting and everyday term to catch this quality of experience of unacquainted stimuli: "Sympathie" (engl. likability, not sympathy) [149]. Examples for assessing Talker Quality between unacquainted people by asking for likability are the freshmen study with all students physically present [17], or a corpus based solely on telephone recording without visual information [318]. Even standardized methods for QoE assessment of communication services include a method of using a single scale [162]. However, other approaches explicitly study various subordinate aspects of QoE or try to disentangle a potentially multidimensional space of QoE features for more diagnostic reasons [216]. This is also approached for Talker Quality in Sect. 2.2. There are other questionnaires available that could be considered. Unfortunately, mostly ad-hoc questionnaires with various designs have been used for this purpose (cf. Sect. 2.3.1).

An exception to this is the *interpersonal attraction scale* [246] that is quite established for applying it to already acquainted relationships. However, it does not seem to be suitable for assessing the first impression in an unmodified version, as it contains 10 items each, for visual physical attraction (like clothing), task attraction (competence and work compatibility), and social attractiveness (which items do seem suitable apart from the item 10: "I sometimes wish I were more like him (her)."). A revised version is presented on the author' s homepage, providing only the six items with highest α-values.[1] It uses a 7-point Likert answering scale. The final score is stated to be the sum of all six items separately for all three distinct dimensions. Another instrument is the *interpersonal judgment scale* [60], using six items (intelligence, morality, knowledge, adjustment, liking, and collaboration) with a Thurston scale, i.e., for each item seven possible sentences, indicating various levels of, e.g., liking, are presented, from which the best fitting one has to be checked. The seven levels correspond to values of 1–7. Only two items are used to calculate an overall value by summing the point, as the other four items are distraction. Much work on the similarity effect for social attractiveness has been done using this scale and it has been shown to be strongly correlated to the social attraction dimension of the interpersonal attraction scale for established relationships [31]. Both instruments were tested for their applicability to assess Talker Quality in passive scenarios, as presented in the following section.

Despite of these questionnaires, a single scale is reasonable and comes handy, if just the overall attitude is of interest. Prerequisite is, however, sufficient reliability, consistency, and validity. Consistency means here inter-subject correlation, reliability can be tested by repeated tests or inter-experimental comparisons, whereas validity can be checked by comparing single scales with full-fledged questionnaires (internal validity) or by comparing data with results after acquaintance. These criteria are addressed in the next section.

---

[1]http://www.jamescmccroskey.com/measures/attraction_interpersonal.htm.

The current practice in the domain of QoE is the just mentioned method of presenting stimuli (in a random order) and rating them. While this is an established and meaningful method, this data is usually used for calculating mean opinion scores [162] and testing for differences in means, disregarding variance as noise, while the distribution itself or even the minimum can provide more relevant information considering the aims of assessing QoE [152]. An approach to make better use of variances in such ratings in passive scenarios has been provided by Kenny with the *Social Relations Model* [186]. For the domain of studying interpersonal social attractiveness in humans, he proposes to conduct experiments, where people rate each other, either in a block design or in a round-robin scenario. In a full block design, people are divided into two subgroups, where each person of one subgroup rates all people from the other subgroup. In a round-robin design, each person rates all other persons of that group. In contrast to these reciprocal designs, a half-block design is typically applied in passive scenarios for QoE. In a half-block design, each person of subgroup A rates each member of subgroup B, but not vice versa. Both reciprocal procedures require to record all participants to obtain stimuli and can thus only be used in humans, not for HCI. By applying a half-block design, reciprocity can obviously not be studied. The ratings are then processed with the aim to separate three components from the grand mean of ratings:

1. The *target effect* reflects the tendencies of the stimuli (the rated persons) to be perceived as likable or unlikable (in our case). This is similar to the typical mean opinion score of a stimulus, except that it is normalized by subtracting the grand mean of all ratings.
2. The *perceiver effect* reflects the tendencies of the raters to perceive others more positively or negatively. In conjunction with user traits and states, this information could be analyzed, for example, to identify influencing personality factors and to define rater groups, which might be valuable to consider in HCI.
3. The *relationship effect* specifies the individual rating given by one perceiver to one target when perceiver and target effects are removed. This relationship provides the means to study effects of similarity in certain traits and reciprocity (Sect. 1.3).

A significant effect is tested with t-tests against zero, by calculating the variances as random effects (Appendix B in [186]). Due to these advantages of this approach, the round-robin design was applied to study more closely personality effects on Talker Quality (Sect. 2.3.2).

While single rating scales and validated questionnaires provide meaningful instruments to assess well-established concepts, such as global Talker Quality or certain aspects of QoE, there is the need to identify perceptual dimensions in the first place, as represented in the lens model (Sect. 1.2). For this aim, two different approaches are established in acoustic research: One the one hand, full questionnaires are developed and applied, which address in particular subordinated aspects of a global concept, as the exact number and mappings from sets of items are not yet known. Data from the individual items is reduced in dimensionality by data reorganization based on correlation and covariance, and by selection of only the (most) relevant representations. As one method, factor analysis treats the data as sample of a larger population and therefore models latent variables by including an error in the model. As such latent variables are expected to be not independent in cognition, non-orthogonal rotations are often applied [39, 71]. As an alternative to a factor analysis, principal component analysis does not include error variance, but models all variance. The selection and revision of items and labels might be conducted by applying qualitative methods, such as focus groups and repertory grids (see below). In the end, a smaller number of factors or components are obtained that is representing most of the variance in data. These factors or components can be interpreted and labeled by evaluating the degree individual items contribute to them.

One the other hand, instead of rating each stimulus separately, overall distances can be assessed by directly comparing two or more stimuli. The answer for each trial can either be directly a metric value on a distance scale or a binary response of similarity or dissimilarity. Such binary responses can be transformed to distances, for example, by counting them for each pair. Once, a distance matrix is obtained, the potential maximum of $n - 1$ dimensional space is reduced to a manageable number by multidimensional scaling [38]. The advantage of assessing (dis-)similarities between a few stimuli, typically two or three, is the missing label that a questionnaires rating scale inherits, along with the issues of properly and consistently addressing the target concept even in the perspective and understanding of the participants. By multiple comparisons all relevant perceptual dimensions to separate the stimuli should be reflected in the multidimensional distance measure, as long as they are not masked by more salient dimensions. In contrast, a questionnaire can only assess the dimensions depicted and thus depend on the design. However, questionnaires are more time effective and provide, due to the labels, (hopefully) well-defined instructions to the listeners and a good basis to identify the dimensions obtained. Therefore, both methods, questionnaires and direct comparisons were applied to study perceptual dimensions of voice and speaking style (Sect. 2.2.2).

Apart from these explicit methods of ratings scales, questionnaires, or (dis-)similarity judgments, there are also implicit methods suitable for assessing Talker Quality. The aim is, as introduced in Sect. 1.2, to capture the aspect of the attitude towards the other (artificial) speaker that a person is not consciously aware of, or at least reluctant to report. A method that crosses the bridge between explicit and implicit Talker Quality is the repertory grid technique that has been developed to study personal constructs [183]. It has been applied to the domains of human stuttering [118], but also sound perception [30], and user interface evaluation [145]. Its original aim is to identify elements of a topic and associated constructs on a personal level, e.g., important people/roles as elements to a person, and characteristics to these elements as personal constructs. For the case of Talker Quality of unacquainted persons, the elements are usually not negotiated (as in Sect. 2.2.2), although they might, and the constructs can be perceptual dimensions or speaker attributions.

Basically, a repertory grid consists of two empirical phases, an elicitation phase and a rating phase. In the elicitation phase, stimuli (i.e., elements) are selected and presented typically in triples. For each set of three stimuli, (dis-)similarity is rated and the character of the (dis)-similarity is described by a set of labels. With triples, two labels can be used to describe an individual scale to separate the two similar stimuli from the third dissimilar one. In contrast to established, sometime even very sophisticated methods in sound perception, for example, speech transmission or speech synthesis quality assessment with direct comparisons, such as MUSHRA [160] or double-blind triple-stimulus with hidden reference [159], this phase comprises an interactive, qualitative part in order to select valid elements for the participant, and to elicit and understand the labels. In the own contributions, this discussion of the labels is moved to immediately after this phase in order to follow a more strict procedure (Sect. 2.2.2). As it was applied to identify perceptual dimensions that are common to many people, an in-depth conversation was not considered necessary, as it would have been in the case of assessing attributions. During the second phase, each stimulus is rated on each of the developed individual constructs, resulting, for example, in a three-dimensional grid (elements/stimuli, constructs/scales, ratings) or simply in a filled-in questionnaire. The quantitative data of the ratings phase can be analyzed in various ways, for example, by means of a principal component analysis. For the application to identify perceptual dimension of Talker Quality, the elicitation phase is more important. While, of course, the repertory grid technique still is explicit in its rating and labeling, the interactive, qualitative part, and especially the indirect approach of using (dis-)similarities for eliciting constructs are a step towards uncovering implicit aspects of Talker Quality.

Truly implicit methods refrain from asking ratings or labels from participants. Instead, reaction times (affective priming task [101] or implicit association test [132]) or behavioral measures (e.g., the number of ball tosses to different people in a game [204]) are used to infer attitudes, or physiological measures are applied. The implicit association test is a widely used method to study, e.g., valence or stereotypes, and it uses relative reaction times to estimate the cognitive strength of association between two concepts [132]. In a series of originally five, now seven, blocks, participants have to categorize words (or other stimuli) on a computer to either a pair of target attributes or target concepts (words) as quickly as possible by pressing appropriate buttons. For the example of age, words as "joy", "agony" appear in the center of the screen and have to be assigned to the attributes "good–bad" which are located to the upper left and right of the screen. For the concepts, pictures of old and young faces appear in the center of the screen and have to be assigned to the labels of "old–young" [269]. Apart from training blocks, there are two main block to collect the data, where all stimuli are presented in a mixed fashion, and the category word for the attributes and concepts is located together, either "good" with "old" and "bad" with "young" or vice versa. Depending on the hypothesis, one of the blocks is called congruent condition, the other incongruent. The relative difference in reaction times between these two blocks is used to infer the association strength between the concept and the attributes; for the example case, testing the association strength that indicates a preference of young over old people by an association between "good" and "young." A proposed way of analysis is the D-measure that is the within-subject scaled value of the effect. From the congruent and incongruent block, it takes the difference of the mean latencies, divided by the overall standard deviation of the two blocks, which is analogous to Cohen's d for effect sizes [133].

A corresponding version for auditory stimuli exists, where the stimuli for construct assignments, or both assignments tasks are exchanged. It was recently developed and therefore lacks currently the academic popularity. One reason might be the longer stimulus duration required in acoustics (see Sect. 1.4) in order to identify the concept. Especially for speech-encoded aspects, such as social group, the duration required might be even longer, and depended on phonetic realization, which might impede preparation of material for valid experiments. Nevertheless, in one of the first validation studies, the auditory implicit association test (aIAT) shows comparable results, but overall more errors and higher response times than comparable version of IAT using visual stimuli [357]. The stimuli used were in this series of studies: Short animal sounds and spoken animal names studying preferences, spoken words for testing for gender stereotypes, and short greeting phrases for assessing ethnic stereotypes. Other studies confirm the congruency of results to visual IATs [147], e.g., for spoken and synthesized two-word terms and names [255].

In sum, this method seems valuable to confirm results for Talker Quality from explicit methods. It has therefore been applied to a small data set in the frame of a bachelor thesis [121]. Ten female recordings of the short German sentence "Die Sonne lacht.", engl. "The sun laughs" (i.e., "is shining") [27] were selected as the five most likable and least likable recordings. This selection was confirmed in a pre-test with 10 participants applying a forced-choice paradigm with over 90% consistent classification. For the main experiment, 15 new and female-only participants were recruited (aged 21–62, M = 27.5). The main aIAT used "likable–unlikable" as target concept, and "positive–negative" as target attribute. The text stimuli for the attribute assignment were translated from English [255] to German. The D-score of 0.49 (SD = 0.35) is greater than zero ($t_{19} = 6.33$, $p < 0.001$) showing that response times are significantly quicker in the congruent than in the incongruent condition. This IAT effect supports an association of valence with likable voices, which means a positive attitude towards the likability concept activated by likable (female) same-sex voices. It remains open, how this method can be applied for more detailed research topics, such as evaluating individual stimuli. While some debate is ongoing, how relevant the individual stimuli are for the results of an IAT [207], the ranking of the reaction times for the 10 female stimuli is correlated to the ranking of Talker Quality averaged scores from a questionnaire that was filled-in at the end of the aIAT experiment ($\rho = 0.90$, $p < 0.001$), but only, if considering all 10 stimuli. Further theoretical and practical validation is required to decide on the applicability of the aIAT for those evaluations. In order to verify the impact of perceptual dimensions on Talker Quality, the concept (likability) would have to be exchanged with the appropriate percepts, e.g., "dark–bright," and respective stimuli presented.

An alternative to the reaction time approach is to measure physiological responses to unacquainted voices directly. While a multitude of measures for skin resistance, muscle activity of valence-related facial expressions, or pupil dilation are potentially useful [207, 313], due to the lack of application in this book, only neuro-physiological aspects will be, very briefly, mentioned. The before-mentioned issue of potential inequality of phonetic material and, in particular, of a principally longer stimulus duration is an argument against applying the rather well-established event-related-potentials (ERP). ERPs are measured as electronic activity at the scalp by means of electroencephalography (EEG). For attitudes [61], a classical approach is to assess ERP as direct response to a target stimulus in a series of standard stimuli (oddball paradigm). For salient target stimuli, a stronger positive response, especially for electrodes placed at the parietal lobe, can be observed about 250–500 ms after the stimulus (onset), which is why this measure is called P300. ERPs have already been applied to study QoE of audio transmission services [10]. A special and elaborate procedure is required for EEG experiments in order to reduce noise by averaging over multiple instances.

Because of this averaging and the desired longer stimulus durations, the temporally critical nature of ERPs motivates to search for other EEG methods, such as analyzing longer stretches of neural responses. There are initial results that such continuous EEGs represent a valid method for assessing emotions, facial preferences, speech intelligibility, and even speech transmission quality, but a validated method to assess vocal likability in this manner is still missing. The aim is to define measures based on longer EEG signals, for example, the whole duration between two stimuli. An obvious step is to filter the signals in the frequency domain [10]. With the insight into neural asymmetries for attitudes [75] or emotions [68], asymmetries, such as the frontal mismatch in the alpha band (about 8–12 Hz) are considered promising measures [10]. However, it is apparently still highly debated, whether such frontal asymmetry reflects valence or arousal-related emotional states [84, 140, 328]. Nevertheless, an experimental method, material, and evaluation scripts have already been prepared for addressing the topic Talker Quality with continuous EEG in more detail [242].

2.1.2 Own Contribution to Methods and Instruments for Passive Scenarios

By evaluating quantitative user responses that have been obtained in passive rating tests, psychometric properties were studied for instrumentally assessing Talker Quality. Some of these tests have been conducted specifically for evaluating the instruments, while others were analyzed to identify acoustic correlates of Talker Quality (Sect. 2.2.2).

For natural voice recordings, the single German antonym scale for *likable–unlikable* ("sympathisch–unsympathisch") shows sufficient consistency as expressed with the intra-class-correlation (ICC) of $ICC_{(3,2)} = 0.93$ (10 speakers, 20 raters [387]) or $ICC_{(3,2)} = 0.76$ (800 speakers of telephone recordings, 32 raters [54]). Despite the lacking of an analysis of the reaction times, the usage of the scales in both experiments was, anecdotally, easy and quick. The lower consistency for the bigger database is attributed to the varying and, on average, very poor signal quality due to telephone transmission. For the small study of 10 speakers, a new set of ratings was obtained from a different group of participants by applying a longer questionnaire, actually the first version of the speaker-attribution and evaluation questionnaire mentioned in Sect. 2.2.2 [390]. The correlation between the questionnaire factor "likability" (composed of items addressing likability, pleasantness, friendliness, and sympathy) and the single item (likability) from this questionnaire is $r = 0.996$ (46 raters [390]), indicating high validity of the single item. Reliability was tested for the two data sets for the 10 speakers by correlating the factor ratings, respectively, the single likability item from it, with the first ratings mentioned in the beginning of this paragraph [54], and yielded correlations of $r = 0.86$ and $r = 0.88$.

For more recent experiments with new recordings, the ICC for selected parts of spontaneous speech is $ICC_{(3,2)} = 0.85$, and the correlation of this item with the factor of social attractiveness is $r = 0.95$ (15 speakers, 33 raters [110]). Ratings from the single item, or the factor, correlate with ratings obtained from a single continuous scale in another experiment ($r = 0.89$ and $r = 0.87$, respectively), in which the same recordings were used (15 speakers, 29 raters [108]). In conclusion of these experiments applying a passive scenario, inter-subject consistency within an experiment, repeated test reliability between experiments, and internal validity for questionnaires can be considered as sufficient for the single German antonym scale.

Although a single item is handy, and the speaker state and trait questionnaire, which has been developed in parallel to applying the single scale, has not proven to be superior for the sole aim of assessing Talker Quality, the other established questionnaire has to be considered as well.

To evaluate these two questionnaires compared to the single, validated bipolar scale "likable–unlikable," the 10 speakers of the EmoDB, for which liking ratings already exist [387], have been rated again in a new study: 15 listeners used the social attractiveness part of the interpersonal attraction scale, and 15 different listeners used the interpersonal judgment scale. The ratings of both questionnaires averaged over participants do not correlate with each other ($r = 0.46$, $t_8 = 1.47$, $p = 0.18$). However, the single item of likability does correlate with both instruments, the interpersonal judgment scale ($r = 0.90$) and the interpersonal attraction scale ($r = 0.71$). With this additional evidence of validity for the single scale of likability, this bipolar scale seems to be sufficient, as long as an overall score of social attractiveness is of interest. Along with the lower correlation with likability, the interpersonal attraction scale also exhibits an unsatisfying ICC ($ICC_{(3,2)} = 0.55$), so that the interpersonal judgment scale ($ICC_{(3,2)} = 0.87$) is advised for the case of unacquainted voices in the case of not using a single scale. However, only the social attractiveness part of the interpersonal attraction scale was used, so the validity of the whole questionnaire including, e.g., competence, could not be tested and a more comprehensive repetition of this validation should be conducted to confirm the results.

2.2 Perceptual Dimension of Voices[2]

The search for perceptual dimensions aims at underlying, and often assumed implicit, perceptions that constitute human categorization and evaluation. According to the lens model, these dimensions or characteristics are called proximal percepts (Fig. 1.1) and they constitute evidence from the world that is only used in a later stage for interpretation, attribution, and evaluation. Therefore, asking people for such perceptual categories is not a valid procedure, but dedicated methods are applied to identify them. For the acoustic domain, several major proximal percepts or "auditory qualities" [342] have already been identified for stationary sounds. These separate different kinds of sounds, noises, and even musical instruments, such as *pitch, loudness, sharpness, roughness, duration*, or *rhythm* [256, 419]. Zwicker [419, Sect. 9.4] provides initial results for modeling sound pleasantness based on *loudness, roughness, sharpness*, and *tonality*.

For the special domain of degradations in speech transmission quality, four dimensions were identified for the passive listening situation. These are *coloration, discontinuity, noisiness*, and *sub-optimal loudness* [371]. Quality of experience assessed as a global concept in the listening phase of a telephone call can be modeled by a linear combination of these percepts.

2.2.1 State of the Art in Perceptual Dimension of Voices

In principle, the same results and methodological approaches can be utilized for perceptual dimension of Talker Quality as well. However, the more fascinating research questions are related to specificities of human speech perception, and these are not only related to pitch and timbre, but also articulation.

This holds also for the domain of synthetic speech. Historically, intelligibility was the first challenge for this domain, but it can by now be considered as solved issue. Nevertheless, with technological advances the interest in perceptual dimensions of Talker Quality has even increased. Meanwhile, a traditional pair comprising intelligibility and naturalness have reached an adequate level, this interest has been expanded to perceptual dimensions not necessarily emerging only from technical degradations. An overview summarizes several studies resulting in the five auditory qualities *naturalness, prosodic quality, fluency and intelligibility, absence of disturbances*, and *calmness* [151]. Apart from the one dimension *absence of disturbances* that is obviously only determined by technically induced degradations, calmness, naturalness, prosodic quality, and fluency are also found for natural speech, as a result of own contributions (Sect. 2.2.2). The questionnaire used for the studies on synthetic speech is based on the development for natural voices

---

[2]Parts of this section have been published previously [386, 394].

as outlined in the next section (Appendix A.1). Obtaining a similar factor structure indicates an increasing assimilation with perceptual dimensions for real speakers. This has been expected, given the increasing naturalness due to non-uniform unit-selection, deep-neural networks, and hybrid synthesis approaches. Unfortunately, the dimensions for synthetic speech have yet to be thoroughly tested on their impact on Talker Quality. Currently understudied domains, like audio book genre and/or children end users [190] establish, of course, further challenges and maybe additional perceptual dimensions.

In order to study these, however, apparently strong human-specific social attributions and evaluations have to be controlled for, as they tend to mask descriptions of voice and speaking style. If the human sound source is not disguised, human listeners attribute, among others,

- speech pathologies [205, 259],
- sex [258, 259],
- age group [41, 229, 409],
- and regional and social background [312].

Even when excluding speech pathologies and recruiting speakers similar in sex, age group, and background. empirical work has come to mixed results. Table 2.1 provides a summary, where identified dimensions from various studies are categorized. This includes dimensions reported by Scherer, who revised the lens model to speech-based attributions [309]. According to this summary, the prosodic features of average pitch and tempo seem to represent established perceptual concepts. Other frequently occurring dimensions are related to the categories pronunciation (articulatory precision), vocal effort, intonation (melodiousness/activity), voice quality (laryngeal settings), and timbre.

Table 2.1 Dimensions found in related work and mapped to categories

Categories	Terms used for dimensions/factors
Average pitch	*pitch*[98, 258, 267, 309, 325], *masculinity*[a][206]
Intonation	*melodiousness*[98], *variability*[206], *sloppy*[309]
Rhythm	*rhythm*[309], *regularity*[309], *fluent*
Tempo	*tempo*[98], *duration*[258, 325], *rate*[309], *animation*[362]
Pronunciation	*articulatory quality*[98], *precision*[309], *clarity*[362]
Timbre	*voice quality*[b][98], *magnitude*[362], *depth, warmth*
	thinness, sharpness, gloom, flatness, dryness, nasality[309]
Voice quality	*creakiness*[206], *hoarseness*[258, 309]
	resonance, roughness, breathiness[309]
Vocal Effort	*effort*[258, 325], *harshness*[309], *loudness*[309]
Others	*affectation*[309]

[a]*Masculinity* may also be assigned to effort or timbre
[b]This dimension reflects "brightness" and is therefore better assigned to timbre instead of voice quality

It becomes obvious from this table that there is a mix in complexity of the dimensions: For example, the first dimension found in [206] is named *masculinity*, and is actually an evaluation related to "relaxed," "dark," and "low-pitched" voices in male speakers. Another example for such an evaluative dimension is Scherer's first factor pleasantness including "clear" and "melodic," although there are separate factors of *precision* and *sloppiness* [309]. In contrast, timbre is apparently a superordinate category for multiple perceptual impressions.

The important characteristic of this approach to perceptual dimensions is its status representing a non-expert, "average" listener. Other, expert driven taxonomies exist, typically for pathological voices. An expert scheme provided for normal voices (but being inspired as well by approaches for pathological ones) is the one by Bose [40], mostly applying five levels of categorization. It separates pitch (and its variation), loudness (and its variation), timbre (sonority, darkness/brightness, faucal narrowing, voicing attack, source of noise, gentle-powerful voice, and its variation), tempo (and its variation), articulation (precision, superlaryngeal settings), and more complex percepts (voice quality settings, rhythm, accentuation, tension, and its variation). This taxonomy is quite similar to the above-mentioned categories. However, it covers articulatory specialties that might not be describable by a majority of ordinary listeners, such as faucal narrowing or super- and laryngeal settings. The most obvious contrast to non-expert listeners, however, may lie in the difficulty to notice and report exact levels of each category for each voice: Where experts can provide a diagnostic evaluation, apparently, only the most salient features are recognized and noted on a questionnaire by ordinary people, but not small amounts of nasality or lip protrusion.

2.2.2 Own Contributions to Perceptual Dimension of Voices

The lens model separates a perceptual layer from the subsequent attribution of a person-related concept. Based on this separation, two questionnaires were developed to cover (a) perceptual dimensions of voice and speaking style (these are the proximal percepts of the lens model), and (b) voice-driven person attributions of states and traits, with the goal to study relations between both stages to social attractiveness which is identified as Talker Quality (quality of experiencing a human-like voice in Sect. 1.1). In order to develop these two questionnaires, a series of experiments was conducted for German, applying both approaches presented in Sect. 2.1.1, full questionnaires with subsequent factor analysis, and direct comparisons to estimate distances that are reduced in dimensionality by multidimensional scaling.

Based on two focus groups with altogether 10 colleagues from speech science and psychology, a prepared list of items to describe voice and speaking style, based on literature from Table 2.1, was discussed, enhanced, and revised. The resulting two questionnaires were validated for a first time with a listening and rating test [390]. Five male and five female speakers were selected from the

Berlin EMOdb [55] uttering 10, semantically and emotionally neutral sentences. 46 participants (25 females and 21 males, aged 21–37, M = 27.1, SD = 4.87) were asked to describe the voice and speaking style, i.e., the perceptual dimensions, on 25 antonym pairs. On 13 antonyms, the attribution of speaker states and traits is assessed. The rating scale was a 7-point scale for each antonym. Each participant was presented 20 stimuli, which means two sentences from each speaker, using Sennheiser HD 485 headphones. The subsequent factor analysis resulted in the dimensions *relaxed–tensed, dynamic–monotone, dark–bright, professional–unprofessional, fluent–disfluent*, and *voluminous–thin*. Only the first factor, *relaxed*, correlates with the person-directed scale of social attractiveness from the other part of the questionnaire ($r = 0.78$).

In contrast to the very detailed descriptors of voice quality (Table 2.1) found by Scherer [309], this validation results provide no support for naive listeners to distinguish such articulatorily defined voice qualities in normal speakers. Instead, all the related items, "breathy," "creaky," "hoarse" (and "nasal" that is strictly speaking not a voice quality, but a superlaryngeal setting [213]) load on a single factor. As typical listeners neither have learned precise definitions of labels applied, nor have profound knowledge about articulation, the latter results should not be overestimated, especially in the light of too many studies recruiting early career students which might have already some expertise. Therefore, new items to assess in much more vague terms "remarkable" and "unremarkable" voices were included to further study this perception, instead of asking for expert concepts of articulation. Also, there was an apparent connectivity between a "low" voice (pitch) and a "dark" voice (timbre). Direct-comparison experiments were conducted with the aim to validate these findings and to further improve the questionnaire by better fitting labels.

The method of direct comparison chosen here uses triads of stimuli with the same read sentence from different same-sex speakers. From this triad, participants had to select two stimuli that are more similar compared to the third one, according to "voice and speaking style" [393, 394]. With this rating, the similarity and dissimilarity had to be labeled as well, in order to interpret the dimensions afterwards, but also to collect non-expert's descriptive terms for the questionnaire (Fig. 2.1). They also had to report, whether the labels provided are considered as positive/negative or neutral. Details of the three separate experiments in German, and one in English, including the dimensions identified, are presented in Table 2.3. The first experiment used the same 10 male and female speakers from the last rating experiment. As listeners, also male and female speakers (aged 21–37, M = 25.9, SD = 5.2) were recruited. This experiment was conducted in a complete design, i.e., comparing all 10 speakers of one gender with each other in one block after another, just differentiating the sentence used to avoid boredom. The number of triads was 10, according to this formula: $\frac{n!}{3!\,(n-3)!}$. For the three other experiments, more speakers were chosen, namely 13, each for German males and females, and Australian males. This resulted in the separation by gender, with 15 same-sex listeners each (males: 22–46, M = 29.8, SD = 7.66; females: 21–36, M = 27.3, SD = 4.25, Australian males: 18–44, M = 26.7, SD = 6.39) to potentially improve

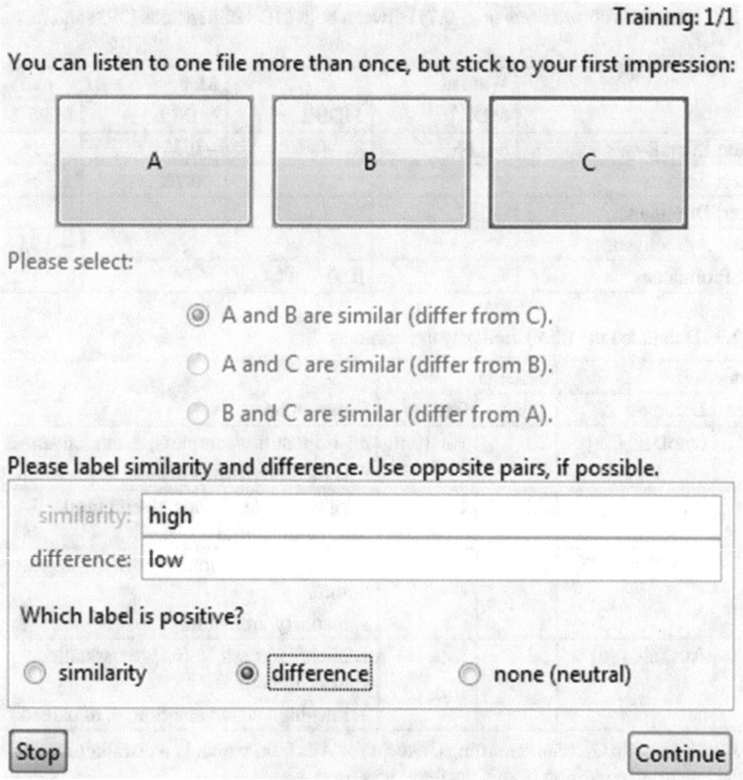

Fig. 2.1   Screenshot of the experimental interface in English [394] (printed with permission)

consensus by reducing the potential elicitation of gender stereotypes and sexual preference. With 13 speakers of one gender instead of five, these two experiments could not be conducted with a complete set of combinations, which is why a balanced, incomplete design was chosen [58] that resulted in 52 instead of 286 triads for 13 speakers. Participants without background in phonetics, acoustics, or the like were recruited. AKG K-601 headphones were used for playback in Germany, and Sennheiser HD 650 in Australia. The number of dissimilarity choices for each pair of speakers was regarded as distance measure and fed into multidimensional analysis.

The first, mixed-gender, experiment confirmed how useful this particular application of direct comparison is in order to elicit labels, and especially to validate the dimension. With only five speakers for each gender, the resulting dimensions are far from exhaustive, of course. Nevertheless, it has to be stated that the questionnaire factors and the dimensions of distance measures are not identical. Table 2.2 gives the correlations between these two approaches for the same 10 speakers, showing results for only one of the seven perceptual dimensions, but for four of the five speaker attributions. This providing initial support that speaker (dis-)similarities are

Table 2.2 Pearson's correlations ($r \geq 0.7$) between both MDS dimensions [393] and factors from [390]

	Women		Men	
	MDS 1	MDS 2	MDS 1	MDS 2
Speaker: Likability	0.96		−0.74	
Speaker: Activity			0.72	
Speaker: Dominance	−0.82			
Speaker: Attractiveness				−0.73
Voice: Proficiency		0.92		

Table 2.3 Details on the (dis-)similarity experiments

Talkers		Listeners		
Gender	Database	Triads	Gender	Dimensions
5m/5f	emoDB[a] [55]	20	10f/10m	m: 1. likability/calmness, 2. attractiveness
				f: 1. likability/conformation, 2. proficiency
13 m	PhonDat 1 [27]	52	15m	1. calm–active, 2. factual–emotional, 3. natural–unnatural
13 f	PhonDat 1 [27]	52	15f	1. tension–relaxation, 2. positive–negative timbre,
				3. maturity–immaturity
13 m	AusTalk [96]	52	15m	1. pitch/darkness, 2. (un)-remarkable voice&timbre,
				3. low–high active emotion, 4. unnamed[b]

All databases provide German recordings, except for AusTalk, which is Australian English
[a]Only the "neutral" condition of this database was used
[b]Despite some bias towards proficiency-related labels, no clear naming was possible

more related to speaker states and traits than descriptors and voice and speaking style, despite the instruction. Consequently, the dimensions were named after the correlating factors (Table 2.3).

The more detailed, gender-separated experiments applying direct comparisons were conducted to further investigate this discrepancy. Instead of two dimensions, both larger data set resulted in three dimensions. In order to characterize each of them, the labels of the six most extreme speakers were chosen. All labels associated with each of the 3×3 pairs of extreme speakers were classified according to the categories of Table 2.1, including others like emotion, and counted. For example, if one speaker pair was labeled as "high–low," the category pitch was increased by one. In parallel, consistency was marked. If, in the end, there were too many opposing labels, e.g., also "low–high," this attribute was neglected for characterizing the dimension. All frequent categories were inspected for consistent labels. This procedure resulted in the naming of dimensions in Table 2.3.

Again, a lot of labels were submitted that attribute speaker traits and states. But apart from *maturity* for females, with its apparent relation to age and stance, and the Australian *emotion*, all other dimensions were dominantly addressing descriptors of voice and speaking style. While no identical dimensions could be expected from sets of 13 speakers, this selection of dimensions that occurred do have similarities, such as *maturity* for females and *pitch/darkness* for Australian males; *calmness* for German males and *tension* for females, or the various *timbre* or *proficiency* related dimensions. The qualitative analysis did not provide sufficient certainty to reflect these similarities in the naming, though. Still, the major result is the mix of factual descriptions, like "high–low" with interpretations, related, for example, to tension. Apparently, the listeners could not refrain from making attributions, and reported much more complex and, in particular, more high-level/interpretative aspects than listed in Table 2.1. This is also reflected in many interesting label pairs, such as "low–bright," "emotional–flat," or "relaxed–high." In conclusion, the direct comparisons confirmed the need to include a relaxation–tension related dimension, and to expect complex dimensions of coupling pitch-brightness to one, and various voice qualities/timbre aspects to another remarkable–unremarkable dimension, confirming initial results from the first questionnaire version [390]. The connectivity between a "low" voice (Pitch) and a "dark" voice (Timbre) can be physiologically grounded in cases of other variables kept constant (e.g., muscular tension): As testosterone (causing male–female differences) increases the size of the larynx and the length of the vocal folds [348], whereas with increasing height, the correlated vocal tract size increases and the formant dispersion decreases [114], there is a relationship between age and gender with formant frequencies and fundamental frequency. In addition, there might be social reasons for speakers to raise pitch and produce a brighter spectrum simultaneously in order to appear more friendly and submissive [92]. Therefore, the very detailed differentiation of aspects for the categories timbre and voice quality (Table 2.1) by Scherer [309] has to be rejected based on the current empirical evidence. For the lens model, the proximal cues for naive listeners consequently have to be considered as less directly related to articulatory cues. Methodologically, expert ratings on specialized schemes may be applied in search for proper acoustic features that resemble distal cues, but apparently, they do not properly reflect naive listeners' perceptual dimensions.

According to these results, the questionnaire was revised to its current version with 34 bipolar items (Appendix A.1). Apart from additional labels to include the new dimensions, some new labels reflect the elicited wording, e.g., "professional–ordinary" for the descriptive part of voice and speaking style. The same 13 German male and 13 German female speakers were presented in a randomized order to same-sex listeners, 31 men (18–65 years, M = 31.2, SD = 10.48) and 30 women (18–39 years, M = 25.9, SD = 5.30) [386]. Two sentences were selected for each group. No listener reported a background related to speech communication (phonetics or the like). For playback, AKG K-601 headphones were used. Horn's parallel analysis and the subsequent factor analysis with oblimin rotation resulted in a very similar factor structure. While the *Tempo* factor was not evident in the male speakers, most likely due to missing tempo differences in the stimuli, all other five factors found are nearly identical. So, apart from the female *Tempo*, with a Cronbach's α of 0.69, there is *intonation* ($\alpha = 0.82$ for males and 0.83 for females), *fluency* ($\alpha = 0.83$, 0.89), *softness* ($\alpha = 0.70$, 0.85), *precision* ($\alpha = 0.81$, 0.69), and *darkness* ($\alpha = 0.74$,

0.83). For the male data, the items *"not breathy," "remarkable," "not nasal," "not creaky"*, and *"clear"* were excluded due to low inter-rater consistency. The remaining items show sufficient consistency, e.g., the average intra-class-correlation (ICC) of these items equals 0.825. For the female data, only "not breathy" is rated inconsistently (ICC = 0.33). The average ICC of all included items is 0.857 for the females. The speakers differ significantly on all factor scores (refer to Tables 2.4 and 2.5).

Just concentrating on perceptual dimensions of voice and speaking style resulted therefore in a confirmation of a combined *pitch/darkness* dimension that is comprising the perception of pitch and a dark–bright timbre, an additional timbre dimension named *softness*, *intonation* that reflects the former factual–emotional aspect, and the proficiency-related *Precision* that is addressing pronunciation accuracy. Inferring from the item loadings, the male aspect of calmness-activity found in the direct-comparison is reflected in *fluency*, although the low tempo-related loadings were rather low. For the females, tension–relaxation labels are found in *softness*. Furthermore, all these dimensions and the remaining *tempo* represent categories from Table 2.1. Therefore, the basic results from the direct-comparison experiments are confirmed: There are perceptual dimensions in naive German listeners, each reflecting tempo, melody (in the table: intonation), fluency (in the table: rhythm), pronunciation, and pitch level. But the following changes have to be applied: vocal effort has to be renamed to *tension–relaxation*, pitch level includes the timbre *darkness–brightness*, and there are no separate dimensions in the categories timbre or voice quality, but instead there has to be assumed an overall *timbre* dimension separate from darkness–brightness that may depend on the speaker selection. This

Table 2.4 ANOVA result for the male data (F- and p-values)

Factor	Speaker $F_{(12,779)}$		Sentence $F_{(1,779)}$		Interaction $F_{(12,779)}$	
Intonation	19.47	$p < 0.001$	5.18	$p < 0.05$	0.95	$p = 0.494$
Fluency	11.88	$p < 0.001$	8.80	$p < 0.01$	10.38	$p < 0.001$
Softness	7.04	$p < 0.001$	0.40	$p = 0.529$	2.38	$p < 0.01$
Precision	6.52	$p < 0.001$	16.01	$p < 0.001$	5.13	$p = 0.001$
Darkness	12.99	$p < 0.001$	7.33	$p < 0.01$	1.80	$p < 0.05$

Table 2.5 ANOVA result for the female data (F- and p-values)

Factor	Speaker $F_{(12,753)}$		Sentence $F_{(1,753)}$		Interaction $F_{(12,753)}$	
Intonation	9.97	$p < 0.001$	9.88	$p < 0.01$	4.27	$p < 0.001$
Fluency	22.42	$p < 0.001$	77.98	$p < 0.001$	10.79	$p < 0.001$
Softness	18.39	$p < 0.001$	0.025	$p = 0.873$	4.05	$p < 0.001$
Precision	21.76	$p < 0.001$	0.18	$p = 0.683$	4.14	$p = 0.001$
Darkness	17.59	$p < 0.001$	117.69	$p < 0.001$	10.70	$p < 0.001$
Tempo	10.05	$p < 0.001$	1.46	$p = 0.228$	2.53	$p < 0.01$

conclusion applies only within the limits of rather homogeneous groups of speakers (gender separated, no pathologies, no famous or highly expert speakers, no strong social and regional background signaled, a young to moderate age group of adults).

The complex character of the dimension hampers an easy mapping to acoustics with the aim to model the relationship between distal, acoustic and proximal, perceptual cues, according to the lens model. More probable will be also a complex relationship that is best attempted with machine learning and much data. As, however, the current data sets comprise only 13 male and 13 female speakers, linear modeling is conducted as a first, hypothesis-based step. In order to do so, the following acoustic features were pre-selected as meaningful and promising. Only automatic measures were taken into account, as these facilitate the application of the factors for studying attribution processes in the future. Parameters associated with each factor are described in the list below. HNR means harmonic to noise ratio of the voiced parts, and LTAS refers to the spectral moments of the long-term-average-spectrum, for example, the LTAS 1st spectral moment is also called center of gravity (CoG). Duration is not including the leading/trailing silence of a stimulus.

Additionally, the following parameters were extracted: the alpha ratio, i.e., the energy difference between the bands 50 Hz..1 kHz, and 1..5 kHz in dB to represent spectral tilt; Hammarberg indexes with energy of the bands in kHz, i.e., ((0..2–2..5)–(2..5–5..8)) related to the tight-breathy contrast, (2..5) related to hyper-hypo functional voice, and (0..2–2..5) related to coarseness [138]; as well as some measures related to syllable nuclei [80], namely estimates of the number of pauses and syllables. For the range of F_0 and intensity, the difference between the 95% and 5% quartile was chosen to exclude outliers. In order to eliminate an effect of sentence, e.g., on duration, acoustic measures were normalized to the mean, separately for both sentence groups. Linear models are conducted on the pre-selected measures, applying step-wise inclusion of predictors pre-selected based on significance ($p \leq 0.10$).

- *Darkness*: F_0 (mean, minimum), LTAS (spectral moments 1–4), alpha
- *Intonation*: intensity (median, range, standard deviation (SD)), F_0 (range, SD, slope), ((0..2 – 2..5) – (2.5 – 5..8))
- *Softness*: intensity (median, range, SD), jitter, shimmer, HNR, alpha, (2..5), (0..2 – 2..5)
- *Fluency*: no. of detected syllables, pauses, pause duration, estimates of articulation rate, speaking rate, syllable duration (all from [80])
- *Precision*: intensity (SD), F_0 (SD), 1st spectral moment, no. of syllables and pauses, syllable duration
- *Tempo* (females only): duration, duration (only voiced parts), estimates of articulation rate, and speaking rate

In the following, linear models are presented, depicting parameter estimates and significance levels. The models are fitted separately for male and female speakers, as similar relations cannot be assumed. They show those parameters of the list above that were actually included. For *darkness*, the female and male model includes pitch and spectral parameters (Table 2.6). Please remind that the values are not standardized, but averaged between sentences, so that the impact of each parameter on the ratings cannot be directly inferred. For female *darkness*, e.g., two parameters correlate negatively, but F_0 median is included positively into the model to compensate for the stronger effects of F_0 minimum and the 1st spectral moment.

Also for *intonation* and *softness* (Tables 2.7 and 2.8), there are significant models with meaningful parameter values for men and women. Interestingly, *intonation* is consistently described by an increased F_0 range. For *softness*, however, there are quite distinct models including intensity (males) or pitch perturbation (females). The jitter values are not comparable to established values for sustained vowels, but a positive impact of jitter averaged for all voiced parts have been reported earlier [388]. Common to men and women is the expected effect of a Hammarberg index, which is negatively related to coarseness (0..2)–(2..5) and, indirectly, positively related to tension (2..5) [138]. Also as expected, *precision* and *fluency* are more difficult to grasp with such basic automatically obtained parameters (Tables 2.9 and 2.10). For *fluency*, automatic syllable nucleus detection [80] works at least for males.

Unexpectedly unsuccessful was the acoustic modeling on *tempo*, despite a rich body of studies on this topic. The simple measure of plain duration is only included for females. A more meaningful measure, based on automatic syllable detection [80] performs not comparable to plain duration, and was not included in the model (Table 2.11). The estimates of this approach show a significant regression with

Table 2.6 Linear models for darkness: females ($p < 0.0001$, $R^2 = 0.703$); males ($p = 0.0199$, $R^2 = 0.466$)

Parameters	Females	Males
F_0 median	0.3658*	n.a.
F_0 minimum	−0.5888***	−0.020806**
LTAS 1st	−0.2685***	0.4268*
LTAS 2nd	n.a.	−0.3216
LTAS 3rd	n.a.	0.5276
LTAS 4th	n.a.	−0.7238*
Alpha	n.a.	−0.4920**

Table 2.7 Linear models for intonation: females ($p = 0.0295$, $R^2 = 0.264$); males ($p < 0.0001$, $R^2 = 0.673$)

Parameters	Females	Males
F_0 SD	n.a.	−0.3167*
F_0 range	0.2762*	0.4586***
Intensity SD	−0.1389	n.a.

Table 2.8 Linear models for softness: females ($p = 0.0002$, $R^2 = 0.589$); males ($p = 0.0006$, $R^2 = 0.535$)

Parameters	Females	Males
Jitter	0.1706*	n.a.
Shimmer	0.1533	n.a.
(0..2)–(2..5)	0.1337	0.1301***
Intensity median	n.a.	−0.2042**
Intensity range	n.a.	−0.7316

Table 2.9 Linear models for fluency: females ($p = 0.0502$, $R^2 = 0.151$); males ($p = 0.0044$, $R^2 = 0.442$)

Parameters	Females	Males
Speaking rate	−0.2361	n.a.
No. pauses	n.a.	0.5846*
Pause duration	n.a.	−0.7781**
Syllable duration	n.a.	−0.1995*

Table 2.10 Linear models for precision: females ($p = 0.0423$, $R^2 = 0.1608$); males ($p =$ n.a., $R^2 =$ n.a.)

Parameters	Females	Males
LTAS 1st	−1.544*	n.a.

Table 2.11 Linear model for tempo: only females ($p < 0.0001$, $R^2 = 0.5222$)

Parameters	Females
Duration	−0.44***

duration, but not with the factor *tempo*. For pronunciation *precision*, parameters aggregating over the whole stimulus length are not appropriate. Although a significant model for females could be found, the 1st central moment, which can be related to segmental precision for consonants [383], is included, but surprisingly with a negative sign. Therefore, this model is disputable and has to be replaced by a more meaningful approach. Segmental analysis, e.g., vowel formants, should be used instead, maybe even including automatic speech recognition.

In summary, the preliminary acoustic modeling supports the validity of the factors obtained from the questionnaire. Further work will have to deal with more sophisticated acoustic parameters for automatically predicting perceptual impressions of precision, fluency, and tempo, in order to subsequently use them for studying vocal attribution processes. As the acoustic model for tempo reveals disappointing performance, a new automatic acoustic classifier was developed that predicts local tempo perception in German, in order to be used for modeling Talker Quality in Sect. 2.3.2. It is based on a German perceptual model of so-called perceived local speaking rate (PLSR), which is a linear combination of syllable rate and phone rate, both represented as continuous, slowly varying signals that are created by taking the reciprocal segment durations and smoothing them with a Hann window of 625 ms for each sample. The linear combination is 8.14 *syllable rate* + 3.31 *phone rate* + 6.07. It allows calculating range and standard deviation of tempo from the local variability over time. The new model automatically predicts PLSR with a recurrent neural network using long short-term memory cells from mel cepstral coefficients [395]. While this acoustic model predicts tempo averaged over utterances for the database in question with high accuracy (Pearson's $r = 0.9$ for data averaged over utterances), further validations are required before its applicability to extract the perceptual dimension of tempo and generalizability for other languages can be ensured. Alternatively, the approach to aggregate automatically extracted features over time by applying a principal component analysis workflow for obtaining mid-level prosodic features, such as tempo, volume, creakiness, and aspects of pitch level and variation [375] has to be tested on its applicability to represent the identified perceptual dimensions and model attributions more appropriately than the current approach. The relevance of these perceptual dimensions for Talker Quality is presented in the following section.

2.3 Talker Quality as a First Impression

This section is about acoustic and articulatory features that correlate with the Talker Quality of human voices in passive scenarios. Some of these studies rely on perceptual descriptors of vocal behaviors, whereas others use acoustic measures of natural of systematically manipulated speech. During the discussion of the state of the art, studies are taken into account that assessed Talker Quality by addressing "social attractiveness" on dedicated questionnaires and items for liking/likability, for example, showing covariation with "pleasantness," which is strictly speaking not an attitude, but an experience report. All of these selected literatures are considered to address an experience-based individual overall evaluation, a.k.a. Talker Quality, in a passive scenario. There are actually only few studies explicitly assessing Talker Quality by liking/likability of a speaker reported from listeners. Instead of a listener's attitude, the majority of results address speaker traits that are considered to be related to Talker Quality, such as benevolence. These studies are taken into account, regardless, in order to increase the number of studies evaluated. This is different for data obtained in interactive situations, with the major example of the area of alignment or phonetic entrainment (see Sect. 3.2), where Talker Quality is explicitly addressed more frequently. However, recent advances in machine learning inspired the rise of classification experiments from passive scenarios, using a vast variety of acoustic parameters to relate to ratings of Talker Quality ([54, 285], and the series of Interspeech paralinguistic challenges, e.g., [318]). Unfortunately, a strong bias towards male speakers is apparent in the cited literature.

2.3.1 State of the Art in Talker Quality

In search of factors affecting social attraction or aversion based on vocal stimuli, purely descriptive and acoustic approaches have been applied.[3] Descriptive methods are similar to the own contributions of assessing impressions from listeners for identifying perceptual dimensions as in the previous subsection. However, the

---

[3] Parts of this and the next section are being published in a modified version [398].

studies presented in this section typically lack the elaborate procedure of instrument development and dimension identification. For example, 1 min recordings of four speakers—presented to the listeners as prospective broadcast speakers—were rated by 96 male undergraduate students [52]. Two questionnaires were used. The one on voice attributes reveals the two factors, pleasantness and intensity. The questionnaire on speaker credibility resulted in three factors, competence, sociability, and extraversion. Regression analysis for the factor sociability (corresponds to benevolence) includes the factor voice pleasantness, in particular the items "variety," "pleasantness," and at last "fluency" as dependent predictors. Of course, there is no explanation of what voice pleasantness actually is. However, variety and fluency are both known perceptual dimensions depicted in Table 2.1 and can be easily related to potential acoustic correlates (e.g., measures of fundamental frequency spreads, and number of pauses, number of hesitations, or number of corrections).

Using the method of randomized splicing (cf. Sect. 2.1), females rated altogether 48 German and American male speakers [310]. Voice attributes were assessed more than once: by acoustic measurements, phonetic description by experts, the speakers themselves and by means of a questionnaire for non-expert listeners. However, only the impressionistic labels are related to the listeners' attitude: Likable voices exhibit a higher pitch range, as well as more loudness variation for German speakers, whereas this loudness variation correlates with ratings of sociable American speakers, but not with likability. 24 of the American speakers were analyzed in more detail: Significant negative correlations of benevolence with perceived sharpness, pitch height, thinness, and a positive one with gloom are reported in [310].

By auditive and manual acoustic analysis, Weirich [382] found for a few selected male speakers a systematic preference of soft vocal attack and breathy voice (for benevolence), and a calm, relaxed voice with precise pronunciation and strong, continuous voicing (for dominance). The questionnaire data on personality attributions were analyzed with a principal component analysis of all 25 speakers rated. As an interesting result, likability loads on both principal components of the questionnaire used, benevolence and competence. Quite similar results are reported for males only, as low pitch, "clear" voice, variable loudness, and a constant voice "capacity" are related to likable voices [189].

These studies show a consistent person evaluation, or even Talker Quality assessment in passive scenarios, which can be related to vocal perceptual dimensions and acoustically stimulated attributions. Combined with the own results from the previous section, Talker Quality was found to be correlated with—and based on the lens model assumed to the dependent on—perceived talker *tempo*, *softness*, *fluency*, *intonation* (comprising variability in pitch and loudness), and *pitch/darkness*. Whether evidence for the further dimensions of *relaxation* and articulatory *precision* are valid has still to be confirmed. In the remainder of this section, a more quantitative approach is presented in order to develop acoustic models of Talker Quality.

Articulation rate is affecting the perception of tempo, and rate is actually one of the best studied acoustic parameters when dealing with listeners' attributions.

Analysis and re-synthesis studies found an ideal-point relation to benevolence (i.e., an inverted U-shaped relation) and a positive correlation with competence [12, 49, 50, 327, 338]. The listeners' own speaking rate moderates this effect by providing an individual reference of a normal rate that the perceived rate of the speaker is evaluated against [104, 338]. For social attractiveness factor scores, a positive correlation with rate was found, including a saturation for the fastest conditions [337]. In conclusion, these results indicate a preference of similarity, which is also known for acquainted people (Sect. 1.3).

Another parameter frequently studied is average fundamental frequency (F_0). At least for male speakers, F_0 correlates negatively with ratings of benevolence, trust, social attractiveness, or pleasantness [12, 49, 51, 67, 131, 382] and competence/dominance [11, 12, 49, 244]. Interestingly, there is an uncommon positive correlation reported between F_0 and likability for German male speakers, not found in US English data, explaining the divergence from other results by a cultural interaction of arousal and dependability [311]. This opposing relationship is also reported for Scottish English data [244].

Still, it has to be pointed out that such results seem only valid for similar situations and similar spoken material: Contrasting results can occur, if listeners are provided with appropriate explanations or context: Slower rates are rated as more benevolent when the speaker is apparently adjusting to context instead of externalizing aspects of personality (in this case slowing down to properly explain something [47]). And a male politician holding a more affective speech (i.e., higher F_0 signaling tension and arousal) is perceived more positively than an opponent with a less adequate amount of activity [273]. Therefore, the question is what attributions are responsible for a positive evaluation of average F_0, for example, body size or tension? It should also be noted that early methods to manipulate F_0 also affect formant frequencies [48, 49] and thus maybe also attributions of body size and perceived darkness of the vocal timbre. The positive correlation with F_0 in Scottish, for example, is based on the single greeting "hello" [244], for which a high pitch might just be socially adequate, and a low pitch inappropriate.

Greater variation in F_0, which corresponds to more variety in intonation, is also rated more benevolent [48, 49, 294] and more competent [294]. Such results confirm qualitative studies reported in the previous section. Also, there is evidence for a positive effect of a rising intonation contour [51, 244, 388]. The last primary prosody, intensity, has shown to be positively correlated with competence judgments, but negatively with benevolence [294]. However, other results indicate a more context dependent relationship [311].

Effects caused by speech manipulation have also shown to be rather independent of each other, as aligned effects show an effect of adding up, while contrasting feature manipulation tend to cancel each other out [294]. Of course, there are also higher level characteristics of voice and speaking style affecting likability, for example, regional accent [124, 388] or disfluencies [247, 388]. But in general, benevolence attribution differs from competence attribution in some acoustic features: especially in rate and loudness, whereas for average F_0, there are conflicting results reported.

Data from the Columbia Games Corpus (from 13 speakers interacting in dyads from the objects game part) are analyzed concerning social behavior [131]. An external, crowd-sourcing platform was used to collect ratings in a passive scenario, whether one speaker is liked by the rater. Acoustic measures are conducted on intonation phrase level and z-normalized for speaker. Liked speakers exhibit significantly lower pitch and pitch range, higher intensity, and lower shimmer compared to disliked speakers. Speaker gender effects are not found for the parameters. Lexical information is also analyzed, but will not be presented here to exclude the field of conversational effects until Sect. 3.2.

Using step-wise regression models for the short stimulus "hello," a likability factor of male voices was described by F_0 and, with a negative sign, the harmonic-to-noise ratio (HNR) [244]. The model for female voices includes HNR (negative sign), a rising intonation contour, and the F_0 range. The contradicting result for males' average pitch may lie in the very special stimulus chosen—a greeting—that may be especially loaded with the social function of signaling friendliness, safety, and readiness for conversation. Aggressiveness was also assessed, and loads negatively on the valence factor including the likability sub-scale. Therefore, higher pitch does not seem to signal activation or arousal, so analyzing longer recordings might turn this result into favoring a lower, relaxed pitch. This interpretation does of course imply a non-existing direct relation between likability and acoustic parameter values as indicated by the lens model and, in conjunction with other context effects, leads to critically regard acoustic models of person attributions that disregard limits of social situation and culture.

A summary of the mentioned results is depicted in Table 2.12. It shows some kind of agreement for the acoustic parameters of primary prosodic features. In particular, there is quite some evidence in favor of a lower average F_0, a moderate of fast rate, and variability. In an older summary of impressionistic studies and those with acoustic measures, focusing on marketing, Ketrow concludes, among others, that variability in pitch, rate, and loudness, the so-called "dynamic deliverable" is, among others, relevant for increased social attractiveness [188]. In general, this conclusion can be supported by the compilation presented here, which also includes more recent studies. Some opposing results found in the table might stem from special social/experimental conditions. Interestingly, all non-significant correlations are from analysis studies with natural occurring parameter values, whereas synthesis resulted in significant effects. The non-significant results might be caused simply by invalid stimulus selection that results in non-sufficient variation of the specific parameter values. For example, a linear effect for syllable rate might still be in line with a similarity/moderate rate dependent on the exact values. For intensity and rate variation, there was no dedicated study with acoustic measures found during the literature review. However, qualitative studies provide arguments for the claim that variability in primary prosodic features might be favored [188]. This gap has been closed with the own contributions reported in the next section.

Table 2.12 Relationships between acoustic features and Talker Quality or benevolence

Parameter	Pos. relation	Neg. relation	Not sign.
Lower average F0	[49, 51] **[387]**m [131] **[384, 388]**mf	[244]m	[89] **[108, 385]**mf **[387]**[244]f
Higher F0 range/variation	[48, 49, 294]m [244]f [89]mf	[131]mf	**[108, 384, 385, 388]**mf[244]m
F0 contour monotone/falling		[51]m [244]f **[388]**mf	**[384]**mf [131]mf (last [1-3]00ms)[244]m
Higher average intensity	[131]mf	[294]m	**[108]**mf
Higher intensity variation	**[108]**f	**[384]**m	**[388]**mf **[384]**f **[108]**m
Higher average rate	**[387, 388]**mf **[384, 385]**f	[294]**[108]**m	[51]m vowels only [131]mf **[384, 385]**m **[108]**f
Moderate/similar rate	[48, 49, 327, 337, 338]m [104]mf		
Tempo variation		**[384]**f	**[384]**m
Harmonic-to-noise ratio		[244]mf	[89]mf [131] **[108, 388]**mf
Lower shimmer	[131]mf **[396]**m		[89]mf [244]mf **[396]**f
Lower jitter		**[388]**mf	[89, 244]mf
Voiced-unvoiced frames ratio			[131]mf
Vowel formant frequencies			[51]m vowels only
Formant dispersion within a vowel			[51]m vowels only [244]mf
Higher spectral tilt/alpha ratio	**[403]**f **[108]**m		[244] **[385, 388]**mf **[108]**f
Lower center of gravity	**[387]**mf **[108]**m		**[385, 388]**mf **[108]**f
Higher spectral skewness	**[387]**f **[385]**f		**[385, 387]**m

Reports on non-significant results may be incomplete. Using recordings of males and females is indicated by m, f, respectively. Own contributions are marked bold

2.3.2 Own Contributions to Voice-Based Talker Quality

The perceptual dimensions identified in Sect. 2.2.2 were also tested for their relevance on Talker Quality with correlation analysis. Ratings for the 26 speakers selected studied for the perceptual dimensions [386] were obtained in a separate experiment with a single sentence that is different to the ones used with the questionnaire, and 30 new different participants. Nevertheless, there are significant positive correlations with a single likability scale from 15 female listeners with *tempo* ($r = 0.78$, $p < 0.01$) and *softness* ($r = 0.63$, $p < 0.05$). For the males, the highest positive correlation is with *softness* ($r = 0.70$, $p < 0.01$), *darkness/pitch*, but it has a negative sign ($r = -0.58$, $p < 0.05$). A positive correlation with *activity* is not significant ($r = 0.48$, $p = 0.09$). Subsequent linear regressions with step-wise inclusion of predictors based on the Akaike information criterion result in significant models: For females, all factors but *darkness* are included ($R^2 = 0.86$, $p < 0.01$). A manual model with just the two most important factors, *softness* and *tempo*, instead of five, is still acceptable ($R^2 = 0.66$, $p < 0.01$). For males, similar regression results in the inclusion of *softness* and *darkness* ($r = 0.62$, $p < 0.01$).

A repetition of this approach was conducted with a new database of 300 speakers, whose creation has not yet been fully completed [111]. Its aim is to support research on attributions and Talker Quality, e.g., on the impact of speech codecs, and already provides first results on speaker trait attributions [110]. The full set of speakers, however, has not yet been rated on the questionnaire of voice descriptions. Meanwhile, there are descriptions available for those 10 female, and those 10 male speakers that are rated as most extreme on Talker Quality. 26 listeners (gender balanced, aged 18–38, M = 26.6, SD = 4.78) filled in the questionnaire. For male speakers, *intonation* ($p < 0.01$) and *darkness* ($p < 0.05$) are significantly different for the two extreme groups on Talker Quality of five speakers each, while *precision* does not reach significance ($p = 0.0556$). For the females, *darkness* ($p < 0.05$), *fluency* ($p < 0.01$), and *precision* ($p < 0.01$) are different for Talker Quality [111]. As a surprise, *darkness* shows a contradicting result for these extreme males and females compared to the studies with acoustic measures of average F_0, as brighter voices were more positively rated, just like the other data set. Auditory analysis, however, indicates a bias for these extreme speakers, as the negative speakers all exhibit outstanding voice qualities and/or signs of being bored/depressed, which is not apparent in the bulk of recordings (see Sect. 2.3.2, page 53 for additional discussion). Here, F_0 does not seem to be a sufficient acoustic representation of this perceptual dimension, as is indicated by the acoustic model of *darkness* for the other data set, where not only F_0, but also spectral energy is perceptually taken into account (Table 2.6). Once, all 300 speakers are described on the questionnaire, this assumed non-linearity has to be tested for and modeled in more detail, and a preference mapping according to the lens model can be conducted and compared to the direct acoustic modeling of Talker Quality. All other results are in line with the ones reported as state of the art, and are considered during the parameter selection of the acoustic analysis presented in the following.

Having established the psychometric quality of the single scale in Sect. 2.1.2, own contributions to identifying acoustic correlates of Talker Quality are presented. There are actually three questions arising when considering the related work in acoustic correlates as summarized in Sect. 2.3.1:

1. Are the results from these, apparently rather few, studies on related concepts reliable also specifically for Talker Quality?
2. Is there a gender difference that has been missed by concentrating on male speakers?
3. Are there also relevant acoustic parameters for secondary prosodic features (beginning of Chap. 1), especially in light of the perceptual dimension that have shown correlations with Talker Quality (e.g., tempo or softness)?

To answer these questions, own studies comprise material from female speakers in addition male ones. Also striking is the obvious lack of thorough experiments on timbre-related aspects. Therefore, more than the typical acoustic parameters of tempo and F_0 have been taken into account. Own studies applied the four different sources of telephone speech [54, 384, 388], standardized read sentences [385, 387, 396], short read paragraphs from a dialog [108], and re-synthesis [403]. Studies aiming on automatic classification are not included here, as they do not provide an easily to interpret relationship between acoustic features and Talker Quality, but only a ranking of feature groups [54, 109, 110, 285, 318].

Starting with the primary prosodies of F_0, duration, and intensity (see the beginning of Chap. 1), a negative correlation was found with average pitch for German male, but also female speakers, further confirming this parameter as relevant for Talker Quality [384, 388]. While there is no opposite correlation found, there are also studies without revealing such an effect [108, 385]. In one analysis, there was only an effect for male, but not for female speakers [387]. Table 2.12 summarizes not only the related work, but also own contributions, which are marked by bold.

In contrast to the effects of average F_0, there was no confirmation found for pitch variation, which should reflect the perceptual dimension of *intonation*, neither for males nor females [108, 384, 385, 388]. However, see page 53 for an analysis on longer stimuli, where pitch variation is relevant for Talker Quality. Concerning the pitch contour, one study supports the related work on a positive effect of a rising contour for males and females [384], whereas another shows no result [388].

Average intensity was not analyzed, as frequently, information on the recordings is missing, even for most of the data sets analyzed as own contribution. Identical recording conditions cannot be ensured, which results in a mixing of articulated loudness and intensity induced by recording level and speaker distance to the microphone. However, relative variability in intensity does have an effect for male [388] and female [108] speakers in the absence of results for the respective other gender. One analysis did not result in an effect at all [384]. Interestingly, rate variation is positively related in males, but negatively in females [388].

Summing up the results in light of the related work, again there are several data sets without effects in naturally occurring variability of the respective acoustic features. While this is not contradicting the relevance of the features per se, it rather illustrates that the human evaluation process, and subsequently any acoustic model, has to rely on a bundle of features when differentiating speakers on Talker Quality: In the own contributions, there are always significant differences in the ratings between speakers, but not each feature is significantly correlated with the ratings. Despite a few contradicting results, there is a majority of studies confirming the positive relationship of Talker Quality with a lower pitch, higher variation in pitch and intensity, a rising pitch contour, and higher articulation rate. These acoustic features relate to the perceptual dimensions of *darkness/pitch*, *intonation*, and *tempo* that have been reported as relevant for Talker Quality in the beginning of this section. In conclusion, the own contributions on Talker Quality confirm the importance of these primary prosodic features explicitly, while they are apparently not exhaustive to model Talker Quality for each of the stimuli. In addition, the contradicting results at hand do indicate the missing insight in limiting factors for these effects, for example, material selection or sampling of speakers. For example, the impact of a raising intonation contour will be dependent on the linguistic material. Striking is the lack of systematical differences for gender, so that the often stated and found result of signaling physical attractiveness by indicating youth and fertility in females with a high pitched and bright voice [102, 103] does not seem to be relevant for the presented data on Talker Quality.

Analysis of electro-glottographic data measured at the vocal folds shows several correlations of the speech recordings with Talker Quality ratings. For example, average pitch, shimmer, but also measures related to the opening phase of the source signal were reported as significant [396]. A linear multiple regression model that includes features from this source signal, such as shimmer, spectral moments, and intensity, performs similar to a model with features from the speech signal for the same recordings [387], exemplifying the relative importance of the source compared to the filter of the vocal tract, and could be related to the perception of *softness* and vocal effort.

Secondary prosodies (cf. Chap. 1), including the just mentioned measures from the voice source, have only recently become a topic to study in more detail. In order to answer the third and last question raised earlier in this section, such parameters have been analyzed as well. The little data at hand show no internal contradiction, but some data sets lack significant effects. For harmonic-to-noise ratio, for example, no effect was found [108, 384]. Lower shimmer is related to Talker Quality in males, but not in females [396]. Also positively related is lower jitter for males and females [384]. Spectral aspects also belong to secondary prosodies: Again, for male and female speakers, a lower center frequency of the energy distribution in a long-term-average spectrum (LTAS)—perceived as dark timbre, along with a higher spectral skewness of the LTAS—is positively correlated with Talker Quality [387]. Again, there is also one study without such a result [384]. In other data sets, such effects were found only for males or females [108, 385].

Current attempts on analyzing the impact of secondary prosodies on liking are twofold: There are approaches analyzing acoustic correlates of physical features, like size of the vocal tract (in terms of formant dispersion of comparable vocal tract configurations), which are known to affect physical attractiveness and thus, liking to some degree (Sect. 1.3). Another approach is to study perceptual dimension (Sect. 2.2), as, for example, brightness (spectral center of gravity (CoG), spectral tilt) or roughness (jitter, shimmer, harmonic-to-noise ratio (HNR)). A measure recently drawing the attention for this domain is the so-called speakers' or actors' formant, a peak in the LTAS (3–4 kHz for males [265] and 4–5 kHz for females [345]). This resonance seems to be caused by an epi-laryngeal narrow and pharyngeal wide configuration that is evident in professional speakers, and it is considered as pleasant also for non-trained speakers [220]. A recent analysis regarding Talker Quality shows a relation for male voices [385] that is even stronger than the typical average F_0. However, this effect was not confirmed by a first attempt of spectral manipulation [180], maybe due to missing representation in other acoustic features that are perceptually relevant for stimulating the acoustic effect of this configuration. In conjunction with results from the study of perceptual dimensions, it seems promising to identify spectral correlates of Talker Quality by testing distal features of relaxed or "resonant" [349] voices. Other, more phonetically defined parameters, such as vowel formant dispersion as measure of articulatory precision, have not yet been analyzed with regard to Talker Quality, simply because they require manual or automatic phonetic analysis for segmental selection.

Summarizing the state of the art and own contributions, selected acoustic parameters have become evident by repetition to be correlates of Talker Quality assessed by "likability" in German, and at least other Germanic languages, such as English. The limitation of this insight is the lack of robustness, which is apparent in failure to replicate each factor's significant impact for each of the repetitions. Of course, the effects are representing averaged data, and thus do not cover all speakers, and might not apply for special social and contextual situations. Any further arguments for generalization, e.g., based on evolutionary factors, are not in scope of this book. Considering the primary prosodies, these parameters are considered to bear valid correlates with high Talker Quality:

- **A lower average F_0 is favored,** presumable due to its signaling of being calm and relaxed. This effect might not include special situations, in which a raised (not high) average F_0 is more adequate, such as politician's arousal or making oneself "small." Vocal markers of physical attractiveness (fitness) might play a role as well (for males), but at least in the German data, a potential pattern signaling fertility by youth with higher average pitch in females was not observed. For females, "tension" and "immaturity" was negatively perceived, which might be the origin of likability correlating with lower F_0.
- **Higher variability in** F_0 is apparently signaling warmth, i.e., interest, friendliness, and inoffensiveness, and it is considered as likable.
- **Increased or moderate articulation rate** is also related to perceived competence, but my have a positive effect on Talker Quality by signaling interest and engagement without (necessarily) a vocal tension.

Secondary prosodies have been analyzed less frequently, and have only resulted in these single effects to be considered as confirmed by repetition.

• **Relatively more spectral energy in lower frequencies.** There are multiple acoustic parameters covering thus effect, spectral tilt, alpha ratio, center of gravity, and even spectral skewness. Apparently, this effect relates to a relaxed voice that is also expressed by a lower average F_0.

All these results have been obtained mostly from stimuli of shorter utterances, i.e., one intonation phrase. A final analysis testing the parameters addressed in this section for longer stimuli is done with a much bigger data set of 300 speakers, including males (126) and females (174), aged 18–35 [111]. However, this time several turns of a pre-scripted "pizza ordering scenario" were played back for evaluation. The parameters tested for are the four effects considered "robust," and in addition intensity variation and tempo variation, which are also analyzed as promising parameters, as well as HNR and shimmer, and the speaker's formant.

The speakers of this database were subjectively checked for neither exhibiting a strong regional nor social accent that deviates from Standard German, nor displaying signs of a cold or voice related sickness or speech disorder. Nevertheless, some speakers do exhibit some regional features and super-segmental abnormal voice qualities. No reported hearing issues, which is relevant for the conversational scenarios and recording instructions by the experimenter. From all available material, the pizza ordering scenario was selected from the semi-spontaneous short conversation tests [163]. It was favored over pre-scripted read speech for its more natural style and wording, which we assume to improve reporting Talker Quality in terms of a listeners' attitude instead of speaking proficiency, which is the major task when reading aloud. When preparing the stimuli for a subsequent listening and rating test, the experimenter who answered the simulated telephone calls of the scenario was removed from the recordings. Based on a questionnaire with 34-items that was developed to assess voice-based personality attributions [110, 390], a minimally revised version was created [111], see Appendix B.1. All recordings of the pizza scenario were rated on this questionnaire by altogether 114 students (44 females), aged 24.5 (SD = 3.4), with about 15 ratings per speaker. More details can be found in the Nautilus Speaker Characteristics (NSC) database documentation [107].

The questionnaire itself is including items to cover major concepts of personality attributions, based on existing instruments for the personality circumflex [407] for the first impression of warmth and agency (see also Sect. 1.2), the OCEAN personality taxonomy [293], the three-dimensional model of meaning with valence, activity, and potency, which is, e.g., used for affect [270], estimation of physical attractiveness that is affecting personality attributions (Sect. 1.3), and frequent attributions observed empirically for unacquainted voices [394] (see also Sect. 2.2.2). The questionnaire also included likability, as this also related to the speakers' person, although not considered an attribution, but rather an evaluative attitude that reflects Talker Quality.

A factor analysis of these items revealed five non-orthogonal dimensions warmth, attractiveness, confidence, compliance, and maturity [109, 111], with an outstanding correlation of the first two of $r = 0.77$. Not only because of this correlation, but also due to the item "likability" correlating with these two dimensions (warmth: 0.87, attractiveness: 0.83), the latter being a known halo effect for the attitude towards a person, these two dimensions are apparently related to the attitude towards speakers. Considering the small number and inconsistent groups of raters, the first principal component of warmth and attractiveness is used to represent Talker Quality instead of the single item of "likability" in order to increase validity. This data is used to identify acoustic correlates of Talker Quality for longer stretches of stimuli.

The test to confirm the nine most important and promising acoustic features, named just above, is conducted by means of a linear bivariate correlations. As alpha level adjustment, the Bonferroni correction was used, resulting in an adjusted p-level of 0.00714. As measure for tempo, the perceptually motivated "perceived local speaking rate" (PLSR) was used [282] (Sect. 2.2.2), as stimulus duration could not be applied with varying linguistic material of the spontaneous utterances. PLSR is used for estimating the mean and also standard variation of the NSC recordings. Instead of the PLSR values, a transformation to the value space of syllables per second is given as sil'/s (PLSR/200 $*$ 14).

Results of this analysis are presented separately for speaker's gender in Tables 2.13 and 2.14. For females, only mean F_0 and F_0 standard deviation (SD) show a positive significant correlation. For males, also F_0 SD, as well as averaged PLSR shows a positive correlation; however, the F_0 average does not reach significance. While increasing variability of F_0 for both speakers sets and increasing rate for males is according to the collected results, typical parameters are not significantly correlated. Actually, average F_0 is even correlated negatively

Table 2.13 Pearson's correlation between selected acoustic features and Talker Quality—females

Feature	r	p-value	t-value(172)	Min–max values
F_0 mean	0.25	**0.0008**	3.42	5.38–7.33 ERB
F_0 SD	0.44	**<0.0001**	6.44	0.52–1.29 ERB
Intensity SD	−0.04	0.5975	−0.53	6.35–10.29 Pa^2/s
Rate mean	0.05	0.4937	0.68	3.14–5.14 sil'/s
Rate SD	0.02	0.7845	0.27	3.00–5.64 sil'/s
CoG	0.05	0.5521	0.60	311–1073 Hz
Speaker's F	−0.10	0.1956	−1.30	1.45–16.04 dB

Table 2.14 Pearson's correlation between selected acoustic features and Talker Quality—males

Feature	r	p-value	t-value(124)	Min–max values
F_0 mean	0.16	0.0688	1.34	2.83–5.21 ERB
F_0 SD	0.52	**<0.0001**	6.89	0.24–1.10 ERB
Intensity SD	−0.07	0.4427	−0.77	6.44–9.68 Pa^2/s
Rate mean	0.30	**0.0006**	3.50	2.66–3.90 sil'/s
Rate SD	0.15	0.0913	1.70	2.72–4.07 sil'/s
CoG	0.18	0.0445	2.03	293–789 Hz
Speaker's F	−0.07	0.4169	−0.81	0.73–14.47 dB

for both, male and female speakers, which is counter-intuitive. The reason for so few and even contradicting results despite the conveniently large number of speakers may, of course, be an artifact of the small number and inconsistently selected raters, as described in [111]. However, a more probable cause of these results is suspected in the much longer stimulus durations. In particular, the strong effect of *intonation* (F_0 SD) may mask smaller, especially timbre-related results. The unexpected positive correlation of Talker Quality with average F_0 is still puzzling, as other work with shorter German stimuli repeatedly gave an opposite effect. A subsequent argument stressing friendly (benevolent, non-threatening) attributions and the earlier mentioned stereotypes for attractive (younger) females and attractive males would be improper given the collected other results. In fact, post-hoc checks for average F_0 for the questionnaire factors called "benevolence" and "attractiveness" confirm the positive correlations for benevolence, but not for attractiveness for neither gender, thus not supporting such an argument with attractiveness. There would still be the context to be considered, as the short sentences may have constructed a different frame of interpretation of the perceptual dimensions as the longer scenario, which might be closer to the one elicited by Scherer, who also found a positive relation for German speakers with average F_0 [311].

For the case of modeling instead of single correlations, the contradicting results for average F_0 are resolved anyway. Until here, available acoustic models that are reported in the referenced papers of own contributions have been ignored for the summarization of results in this section, due to the small number of speakers, and in order to focus on identifying the most reliable acoustic parameters. However, for the NSC corpus, simple linear models are used to describe the results for these 300 speakers, with the aim to exemplify the amount of variance that could be explained for longer stimuli durations and more variable stimulus content. By step-wise inclusion with the Akaike information criterion on the most promising parameters, these two models emerge to describe Talker Quality, which is normalized (Table 2.15).

The positive bivariate correlation of F_0 is extenuated a little for females, while the sign for the male speakers switched to negative. The resulting models have not been further analyzed to find the source of the interrelations between the predictor variables (e.g., being confounded with or a suppressor variable for F_0 SD). In smaller data sets with shorter stimuli, for which linear regression models are available, the signs of the bivariate correlations did not change in regression models [385, 387, 390, 396]. Compare to these, the amount of explained variance is

Table 2.15 Linear models for Talker Quality: females ($p < 0.0001$, $R^2 = 0.206$); males ($p < 0.0001$, $R^2 = 0.345$)

Parameters	Females	Males
Intercept	8.6051***	5.7094**
F_0 mean	0.6480	−1.1345*
F_0 SD	5.1326***	8.7584***
PLSR	n.a.	0.0759**

slightly smaller. As one reason for this, the material is semi-spontaneous and thus not as well comparable between the speakers, as the short and linguistically nearly identical utterances or words typically used in rating tests on Talker Quality are. This might have increased the relevance of F_0 SD. And given the high number of speakers considered, it is rather common that partial correlations, etc., that depend on or are moderated by other predictors, appear. Also, this kind of diverse and longer material might not be adequately represented by overall and averaged acoustic parameters, which have been proven to be meaningful and reliable for shorter and similar utterances. Instead, for such stimuli, other acoustic parameters might affect the perceptual dimensions that have to be more relative to linguistic and articulatory factors, such as specific intonation contours to estimate the pitch level or floor or to extract spectral measures only for well-defined phonetic categories. Finally, this last experiment indicates the limits of the grown body of results in terms of duration and linguistic material. For this kind of longer and more varied material, such parameters grounded more in articulatory settings and perceptual classes have to be defined—the perceptual dimensions identified in Sect. 2.2 represent a big step towards this aim, and the acoustic correlates and models should be improved and tested for the requirements of the new material.

To conclude the modeling of passively obtained averaged Talker Quality with acoustic parameters, a closer look on non-linearities needs to be done. As outlined with the presentation of the lens model in Sect. 1.2, non-linearities can be assumed in the inference process of interpreting perceptual dimensions for attribution. The same holds for the evaluation process during the manifestation of Talker Quality itself. Initial evidence for such kind of non-linearities has been reported for speakers with negative and those with positive Talker Quality [384, 388], indicating a bias on the negatively rated speakers towards the presence of special voice qualities, pronunciation issues, and so on. Positively rated speakers, in contrast, seem to exhibit not only the absence of such markers, but also particularly positive timbre and intonation. Such non-linearities should also be apparent in individual stimuli whose Talker Quality cannot be captured well by the simple regression models. A pure machine learning complex model is not in scope of this book, due to its lack in interpretability, but a simple regression tree on the NSC data gives indeed more insight into the relationship between the selected predictors and Talker Quality.

All seven parameters were used with rpart [290], restricting the model to a depth of only three nodes of the regression tree for inspection. While of course, the tree-models perform better than linear regressions for non-linear relations, it is more interesting to examine the parameters finally chosen than the variance explained (Fig. 2.2). In the figure, nodes show the binary partition of data in percentage, as well as the Talker Quality values of the current node (the first principal component of warmth and attractiveness, see page 52, with values from -5.8 to $+6.0$, mean $= 0.0$). Edges connecting the nodes are labeled with the corresponding parameter and its un-normalized value for the split: F_0 mean (F0) and standard variation (F0.SD) in ERB, PLSR, speaker's formant (SP.F), intensity standard variation (Int.SD), and center of gravity (CoG). Color indicates low Talker Quality (white) to high values (blue).

Both data sets have a first split with intonation variation (F_0 SD), for which higher values are positive. The positive group is again split by F_0 SD for males and females with the same systematic. For females (lower tree in Fig. 2.2), there is a positive impact with lower CoG, but a lower intensity variation, and a higher average F_0. The result for CoG hints to voice quality and/or a darker voice. The approach of regression trees forces binary splits. Still, considering the differences in acoustic parameters relevant for the positive and negative first split on later nodes, the resulting two models do support the results of asymmetries in Talker Quality reported for much smaller data sets. The effect for average $F0$ is similar to the one of the linear model. This one and the sign for intensity variation are not confirming earlier results. For the males (upper tree in the figure), tempo is chosen to split the more negative speakers, while the speaker's formant is only selected to split speakers with rather average Talker Quality. The first positive part of speakers (60%) shows a positive impact of lower mean F_0. To summarize, the spectral parameters play only a minor role in modeling, and only when considering non-linearities with a regression tree, while variability in F_0 represents the dominant predictor for male and female speakers. Still, with the exception of average F_0 and intensity SD in females, other primary prosodic features, i.e., female' CoG, and males' average F_0, tempo, and speakers' formant, also confirm their (minor) relevance in line with the directions as summarized in Table 2.12, albeit governed by other features, especially F_0 SD. The relevance of F_0 SD, especially compared to average F_0 in other studies, may be caused by the increased stimulus range, which might allow intonation to carry more communicative features, of, e.g., signaling friendliness, attention, and others.

Up to now, results on the relation between acoustics and speaker-averaged Talker Quality have been presented in this section, as well as on the identification of perceptual dimensions with an impact on Talker Quality. The final contribution in this section is to go beyond averaging for a "mean opinion score," a typical procedure in the referenced literature and stimulus rating in general [152, 162], by taking into account the listeners individuality and their interaction with the talker's quality. For this, the approach of the social relations model, as described in Sect. 2.1.1 and mentioned in Sect. 1.3 is applied. The aim is to obtain additional insights on the status of Talker Quality ratings. Two round-robin experiments have been conducted, to the best of our knowledge for the first time focusing on speech-only first impressions. Two samples from the Nautilus speaker database [111] were asked to return to the lab for participating in a rating experiment after the recoding session. The aim was to explore effects of the three components of target, perceiver, and individual relation (Sect. 2.1.1). Also, the factor of personality was studied for Talker Quality.

In Experiment A [108], 30 German native speakers from the Nautilus database participated (15f, 15m), with a mean age of 27.2 years (range: 20–34). Except for two people who were a couple (1f,1m), the rest of participants were not acquainted with each other. Mutual ratings of these two were excluded during analysis. Each of the sessions took 1 h to complete and all participants were financially compensated for their participation.

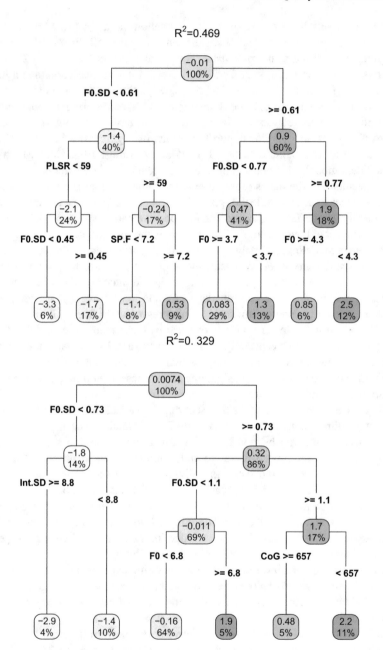

Fig. 2.2 Result of simple regression trees until depths of 3, for males (upper) and females (lower)

Personality was assessed according to the OCEAN taxonomy [245] with self-reports on the short German BFI-10 questionnaire [293]. This questionnaire assesses only the major five superordinate dimensions of the acronym OCEAN, namely openness, conscientiousness, extraversion, agreeableness, and neuroticism. The whole experiment consists of two parts. As the first part is about channel effects and used read speech, only the second part is reported here. All material from one speaker was concatenated with small pauses from the semi-spontaneous pizza ordering scenario of the short conversation tests [163] as stimulus (15–33 s, M = 19.5, SD = 4.3 s) and played back on AKG K-601 headphones. The 29 stimuli were blocked for gender. Participants rated every stimulus on the BFI-10 and on a continuous scale "likable–unlikable" for Talker Quality.

The social relations model analysis was conducted with TripleR [308]. According to Table 2.16, all three effects are significant. The standardized variances show a typical pattern of a comparable amount of perceiver and target variance, but a much higher amount for relationship. In comparison to a German freshmen analysis, where data was obtained from a university's introductory session with all participants were present to see and listen to each target, a similar pattern was obtained [17].

Self-reports on the five superordinate personality traits extraversion, agreeableness, conscientiousness, neuroticism, and openness show no effect on the perceiver component (*What is the personality of people who like others?*), only agreeableness is worth mentioning ($p = 0.97$, $r = 0.31$).[4] This is consistent with a German study of multimodally perceiving universities freshmen [17]. However, there is a target effect from the personality ratings of the stimuli (*Does attributed personality affect likability ratings?*).[5] Assumed agreeable, conscientious, and open speakers are rated more positively. This is a well-known effect for acquainted people [355, 412]. However, for unacquainted persons, the already mentioned study in [17] reports only a positive effect of extraversion on target effects, whereas the other dimensions

Table 2.16 Exp. A: relative variance components of likability

Variance component	Standardized	t-value
Perceiver	0.148	3.384**
Target	0.150	3.389**
Relationship	0.702	20.043***
Perceiver-target covariance	0.020	0.094
Relationship covariance	−0.121	−2.432*

---

[4]Please notice the correction compared to the first analysis [108] that had slightly different results due to an error on preprocessing of the personality data.

[5]Consistent with the lens model, personality effects on the target component were represented by the participants BFI-10 averaged ratings of each stimulus, not the targets' self-ratings, to best capture the raters' information status and attribution processes. Self-reports are not considered relevant here, as such traits first would have to be exhibited in speech, and then perceived by the raters and evaluated with regard to personality. The personality ratings were z-normalized over raters prior analysis.

were not correlated. Still, there is much evidence suggesting a positive effect of perceived benevolence, which is related to agreeableness (Sect. 1.2) on liking and social attractiveness for the acoustic domain and unacquainted people (e.g., [337]).

The relationship component was correlated with dyadic similarity in personality. This similarity was calculated as the absolute difference between a perceiver's self-rating and his/her individual rating of the stimulus for each of the OCEAN traits, multiplied by -1 and then z-normalized. In contrast to [17], which did not find a similarity effect for personality, but only for preferences (clothing, subculture), a significant similarity effect for agreeableness, neuroticism, and openness can be observed from our results (Table 2.17). This indicates that pairs of individuals who are close in these personality traits tend to like each other's voices to a greater extent, even after controlling for their tendencies of "being a liker" and of "being liked" by others.

This statistic analysis allows also for testing for reciprocity, which is typically assumed in longer relationships of acquainted persons (cf. Sect. 1.3). With the data at hand, two kinds of effects can be considered: *Generalized reciprocity* correlates target with perceiver effects, and would indicate that likers are also liked. If present, such an effect might be even negative, e.g., for speed dating, as attractive people tend to be choosy [187]. A correlation with the directed relationship components would be an effect on the individual level of dyads (persons who have individual preferences of a certain voice are also particularly positively rated by that person). In contrast to traits and states, for the concept of liking in acquainted persons, *dyadic reciprocity* is more established than generalized reciprocity [186]. For the Talker Quality, there is also no general effect, but a negative dyadic reciprocity, which might indicate such an asymmetric process for unacquainted people (last two lines in Table 2.16).

In a slightly revised repetition of this experiment, 28 new participants from the Nautilus database were recruited for the round-robin design (aged 20–34, M = 25.86, SD = 4.26, 12 males and 16 females). For this time, nobody reported to know any of the voices or names. In difference to the first time, in Experiment B no personality other-ratings were collected due to time issues. Instead, meta ratings were collected on "How would this speaker rate me?" along with the liking ratings. Therefore, a small rating test was conducted to collect external personality assessments from 10 people separately (gender balanced, aged 18–32, M = 26.2, SD = 4.08).

A comparison with Experiment A shows similar effects of perceiver, target, and relationship effects, albeit the effect size for the perceiver variance is a little

Table 2.17 Exp. A: personality correlations with SRM components	Trait	Perceiver	Target	Relationship
	Extraversion	0.04	0.08	0.01
	Agreeableness	0.31	0.62***	0.12***
	Conscientiousness	0.11	0.50**	0.07
	Neuroticism	0.24	−0.31	0.15***
	Openness	−0.07	0.56**	0.21***

Table 2.18 Exp. B: relative variance components of likability

Variance component	Standardized	t-value
Perceiver	0.082	2.841**
Target	0.165	3.255**
Relationship	0.753	18.773***
Perceiver-target covariance	0.155	0.674
Relationship covariance	0.012	0.217

Table 2.19 Exp. B: personality correlations with SRM components

Trait	Perceiver	Target
Extraversion	−0.42*	0.23
Agreeableness	0.14	0.18
Conscientiousness	−0.02	0.27
Neuroticism	0.07	−0.59**
Openness	−0.05	0.44**

bit smaller (Table 2.18). There is again no generalized reciprocity, and this time even no dyadic reciprocity. While personality cannot be analyzed with respect to relationship due to the missing ratings, perceiver and target effects show significant results (Table 2.19). The effect of openness is confirmed, and the negative relation with neuroticism reaches significance. The other two positive correlations from Exp. A could not repeated. The impact of self-reported extrovert persons to rate more critically is significant in this data set. A similar tendency in [17] was too low ($r = 0.10$).

The social relations model analysis of the subset of Nautilus speakers confirms personality effects in first impressions of Talker Quality, not only for the target effect, i.e., similar to the traditional analysis of ratings in passive scenarios, but also for the individual relationships. As the individual relationship exhibits the highest amount of variance in the ratings, this result is a first step to explain an important variance component for audio-only Talker Quality, by, e.g., personality similarity and negative reciprocity. In the future, aspects of reciprocity have to be studied in more detail, as well as acoustics of the perceiver effect in order to test whether "being a liker" is represented in speech. The relationship component provides an interesting mean to study similarity factors, such as regional background, intonation contours, or (socio-linguistic) idiomatic expressions. A preliminary attempt to test for a similarity effect on averaged acoustic parameters gave no results, and has to be examined with regard to the lens model (Sect. 1.2), i.e., is there a percept of acoustic similarity, or are there only social similarities, such as linguistically signaled regional, age-related, or social similarity?

2.3.3 Own Contributions to Talker Quality of Embodied Conversational Agents

In passive scenarios, audio-visual speech differs from purely acoustic speech signals in the visual impression of the face and the visibly moving articulators. The additional modality enriches not only the acoustic signal for speech perception [249, 287], but also for human audio-visual person attributions and evaluations [7, 63], even for virtual and synthetic agents (Sect. 1.4). However, the audio-visual signal does not change the basic principles of the formation process of person attributions. Of course, some attributions might be more salient in acoustics, whereas others are in the visual domain. For example, acoustic information is referred to transport tension/arousal very well, while the facial information seems to be very salient for positive–negative distinction of emotions, i.e., its valence [136]. Also for other attributions, such as age, gender, sexual attractiveness, or the recognition of smiling (which can actually be interpreted in many ways, e.g., as uncertainty or as happiness), both modalities can be processed in isolation or combination. Nevertheless, there are some facial signals that are apparently neither directly nor indirectly (e.g., by tissue) related to articulation, such as frowning (resulting from contradiction or doubt), eyebrow, and eye movements. Therefore, such signals are independent to information of the acoustic signal, but they may originate in the same cognitive representation. As a consequence, they might have associated, but unique representations in acoustic features, such as a special intonation contour or emphasis strategy for the concept of doubt. Other sources of information, especially those related to non-fronted articulatory gestures or paralinguistic signals of regional and social background, cannot be found in facial expressions, but maybe in other visuals, such as clothing. For social relationship and status, for example, posture and gestures are used for inferencing (Sect. 1.3).

Since the obvious commonalities of visual and acoustic signals originate in basic person features, like personality, relationship, biological age, and fitness, the same attributions and evaluation play a role as for purely acoustic signals. Typical parameters of facial composition are vertical symmetry (especially for sexual attractiveness), hair, skin texture, and descriptors of facial location and movements of organs, such as the shape of the nose, e.g., parameterized by the facial action coding scheme (FACS) and its derived coding features (implemented, e.g., in MPEG-4 [158]).

Two additional factors, however, emerge by the multimodality of audio-visual speech itself. One is the *fit* of those attributions from acoustic and visual information. One example is age, which is difficult to obscure in audio, as well as visual human(-like) speech (e.g., by hair, or skin texture). Thus, information about age should fit to each other on both channels, when synthesizing virtual or truly embodied agents. Temporal *synchrony* between the two signals might also play a role as the other factor.

By playing back short recording of audio-visual stimuli produced by one out of three facial models and by one out of two voice models, the overall Talker Quality, plainly named overall quality in the scale applied, was collected for multimodality. Along with Talker Quality, speech quality and visual quality were also assessed [404], assuming that the participants in such passive scenarios would be able to rate both modalities/modes separately from a multimodal experience. The models are the three-dimensional models "Thinking Head," and "MASSY," and the two-dimensional video morphed "Clone" (Fig. 2.3) as facial models, while the two freely available German voices are the "de2" (di-phone synthesis) and "bits-3" (HMM synthesis). The Thinking Head exhibits a skin texture from photographs of the artist Stelious Arkadiou and the nonverbal signals of occasional smiling, winking, and movements of the head and the eyebrows. Due to the immense technological advances in the last years, these models are not state of the art anymore, which does not reduce the validity of the results, though, as the participants did not have higher experience-based expectations during the experiments. Ten sentences were produced for all six combinations of visual and acoustic model and played back in randomized order to 14 participants (aged 20–32, $M = 27$, $SD = 4.21$, gender balanced). For each stimulus, the participants answered a distraction question on the sentence content, and then rated Talker Quality, speech quality, and visual quality. In a second part, a selection of six sentences was presented in a block for each of the six conditions and rated on a longer questionnaire with antonyms in order to study perceptual dimensions, just like with naturally and purely acoustic speech signals (Sect. 2.2). This questionnaire is based on [2], which aims at assessing attitudes and person attributions specifically for artificial tutoring agents.

Results of the first part show no impact of the sentence or voice on visual quality, while it is different for the facial models. Speech quality is affected by all three variables; however, the effect on speech quality is much smaller for the facial model than the voice model or the sentences. The overall Talker Quality is strongly affected by all three variables (Table 2.20). For the facial models, Thinking Head is better than MASSY, which is better than Clone. Interaction effects between

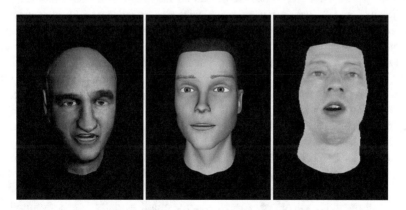

Fig. 2.3 The four facial models, Thinking Head (left), MASSY, and Clone (right). Printed with permission [405]

Table 2.20 Repeated measures Anova results for the first part of the experiment—web version

Scale	Variable	$F_{(\mathbf{df}, 659)}$	p-value	Partial eta^2
Talker Quality	Sentence	(9) = 3.53	<0.001	0.03
	Voice	(1) = 96.71	<0.001	0.10
	Head	(2) = 74.96	<0.001	0.15
Speech quality	Sentence	(9) = 7.19	<0.001	0.06
	Voice	(1) = 249.87	<0.001	0.24
	Head	(2) = 3.03	=0.049	0.01
Visual quality	Sentence	(9) = 0.46	=0.899	0.00
	Voice	(1) = 0.17	=0.679	0.00
	Head	(2) = 389.09	<0.001	0.50

speech and visual quality were not tested due to the small number of participants. The plausibility of results validates this procedure for further application. Finally, at least for this kind of experiment, Talker Quality can be well modeled with a simple linear regression, in which visual and speech quality complement each other, as both are included in a step-wise approach, and as this combination nearly doubles the variance explained (from $R^2 = 0.39$ and $R^2 = 0.34$ for single regressions with speech and visual quality to $R^2 = 59.6$ for the multiple regression, Eq. (2.1)).

$$\text{Talker Quality} = 0.75 + 0.41 \cdot \text{speech quality} + 0.37 \cdot \text{visual quality} \qquad (2.1)$$

For further validation, six sentences with the least divergence in ratings were selected from the first part of the experiment and presented in a similar fashion to a new group of participants via a web-experiment (25 men, 17 women, aged 20–58, M = 26, SD = 5.89). This time, the fit, and synchrony, as introduced in the beginning of this section, were also assessed in addition to the three scales used in the last experiment. The respective questions are *"How well does the voice fit to the head?"* and *"How do you rate the synchrony of voice and lip movements?"*. In order to shorten the web version of this experiment to a maximum of 15 min, the worst performing facial model "Clone" was excluded. The results are depicted in Table 2.21, and show comparable, but not identical effects. Talker Quality is only affected by sentence and the voice, not by the facial model, most likely due to excluding the worst facial model beforehand. In addition, the small effect of the facial model on speech quality is not confirmed, maybe due to the increased number of participants, but more likely due to the removal of the worst facial model that might have decreased the quality of the audio-visual speech, as we did not ask for acoustic speech quality in particular. Based on the multimodal nature of human speech processing, which is an automatic process [249], labeling the item not "speech quality" but something like "acoustic quality" seems to be like asking for the impossible. Therefore, speech quality might actually assess the mode of speech instead of the acoustic modality. Interaction effects between speech and visual quality were not found and thus only the main effects are reported. The two

Table 2.21 Repeated measures Anova results for the first part of the experiment—web version

Scale	Variable	$F_{(df, 999)}$	p-value	Partial eta^2
Talker Quality	Sentence	(5) = 5.49	<0.001	0.02
	Voice	(1) = 155.04	<0.001	0.13
	Head	(1) = 0.87	=0.351	0.00
Speech quality	Sentence	(5)= 8.92	<0.001	0.03
	Voice	(1) = 305.50	<0.001	0.22
	Head	(1) = 0.40	=0.841	0.00
Visual quality	Sentence	(5) = 0.42	=0.836	0.00
	Voice	(1) = 2.87	=0.091	0.00
	Head	(1) = 132.36	<0.001	0.12
Multimodal fit	Sentence	(5) = 1.93	=0.086	0.01
	Voice	(1) = 98.02	<0.001	0.09
	Head	(1) = 0.00	=0.973	0.00
Multimodal synchrony	Sentence	(5) = 1.71	=0.131	0.01
	Voice	(1) = 70.73	<0.001	0.06
	Head	(1) = 78.91	<0.001	0.07

facial models are similar in Talker Quality and do not affect speech quality. Only for visual quality, the Thinking Head with its natural texture is rated more positively. The HMM voice receives better ratings for Talker Quality and speech quality.

Concerning the new scales, multimodal fit and synchrony are not affected by sentence. MASSY, however, is rated to better synchronize with both voices. This is an interesting evaluation result, as the model with the lower facial quality is perceived better for temporal synchrony. The reason is the more precise control of the temporal movements of this model compared to the lower temporal resolution of the Thinking Head model. In addition, the HMM voice shows a better fit and synchrony than the "de" voice. According to the results from [261] for real speech recordings and synthesized speech and facial movements, an interaction effect was expected showing a better fit between the more artificial models (MASSY and de2) and the more human-like models (Thinking Head and hmm-bits3). This, however, was not supported. Instead, for this limited number of versions, the best models give the best fit, limiting the results by [261] to the combinations of completely human and synthetic models for now. According to the smaller laboratory experiment, a linear model was fitted, which results in the inclusion of all four scales and a strongest impact of speech quality (Eq. (2.2)).

$$\text{Talker Quality} = 0.04 + 0.39 \cdot \text{speech quality} + 0.27 \cdot \text{visual quality}$$

$$+ 0.12 \cdot \text{fit} + 0.19 \cdot \text{synchrony} \tag{2.2}$$

In order to analyze the longer questionnaire used for assessment in the second part of both experiments, Horn's parallel analysis was applied to determine the number of factors for the subsequent principal axis factor analysis. For the first experiment, there are three factors found, entitled as *naturalness*, *friendliness*, and *attractiveness* [404], which all are positively correlated with Talker Quality. This quote from page 484 of the article concludes the analysis of the questionnaire:

> TH is considered as friendly as MS, but more natural and more attractive. It is interesting to see that friendliness does not correspond to human-like texture, or natural extra-linguistic movements. We can only speculate that friendliness might depend on other features representing a different personality (such as head shape or more constant slight smiling) rather than on the degree of artificiality.

In contrast to the laboratory test, the second part of the web-experiment was conducted with a between-subject design. 36 participants completed the longer questionnaire. With this bigger data set, a factor analysis is less inappropriate. An analysis similar to the laboratory experiment reveals three factors as well. Based on the loadings of the items, these factors are named *likability*, i.e., Talker Quality, *naturalness*, and *stimulation*. This is in line with comprehensive analyses conducted for more data sets, including interactive studies with my colleagues [172] (pp. 165–167). This questionnaire was revised based on several additional studies, including the ones presented in [172], and resulted in the *conversational agent scale* [379], in which the third factor is called *entertainment* instead of *stimulation*, but based on similar item loadings. Again, all three factors correlate with Talker Quality. As an elaborate analysis of the questionnaire has been conducted for a bigger database of studies, including those not covered with the contributions of this book, please refer to [379] for details. For now, all six scales show high reliability and consistency.

As virtual agents and social robots are dominantly evaluated in interactive scenarios, further results on Talker Quality of anthropomorphic agents, in particular the direct continuation of the studies presented here in interaction, will be presented in Sect. 3.3. This includes the remaining three factors of the full *conversational agent scale* that are related to the agent's wording and ease of interaction.

2.4   Conclusion and Future Directions

The presented work of identifying perceptual dimensions and attributions provides a systematic approach to modeling Talker Quality in passive scenarios. Repeated results show that Talker Quality is related in particular to the perceptual dimensions of *softness* and *intonation* for female speakers, and *softness* and *darkness* for male speakers. The dimension *intonation* seems to be closely related to the attribution of activity–calmness, but additional studies on local phenomena, e.g., on rhythm, might clarify this. Currently, there is still too little data for building a full multi-level model that incorporates person traits and states for Talker Quality as conceptually defined with the lens model, in particular for the relations between perceptual dimensions and attributions. The attribution of being calm is also apparent in synthesized speech, so with increasing naturalness, more of such talker states and traits will be addressed with speech-based services and products.

The extended body of results, as presented in Table 2.12 and with the NSC data, confirms positive effects of increased variation in F_0, articulation rate, and a lower center of gravity. Specifically for males, lower average F_0 is a robust result, while a higher speaker's formant intensity gives promising indication of being a correlate of Taker Quality. The contradicting effect found for average F_0 in females of the NSC data has to be analyzed in more detail, once perceptual dimensions are collected for all speakers: The negative relation found with decreasing values might be a result of an interplay of it with increased F_0 SD and CoG, detectable by inspecting data on darkness.

Along with the confirmed relevance of, in particular, pitch variation, simple acoustic models have been presented as starting point for building such a multi-level description of the formation of Talker Quality. Such a comprehensive model would allow capturing non-linear systematics, which have been found for speakers with negative and positive Talker Quality. However, more complex acoustic parameters are required to take into account more articulatory grounded aspects of timbre and local communicative signals, such as realization of stress and prominence. Bringing together the various levels (for example, in the frame of the lens model) and methods is the requirement for taking the next step in the performance of such models, as has been exemplified for individual relationship effects that are affected by personality similarity. Such an increase in explained sources of variance is required to use the models for predicting individual Talker Quality for smaller listener groups. For archiving this, however, more results on voice- and speech-based similarity are required. In particular, the aspect of social and regional background has been not in scope of this book, but does play a role in individual Talker Quality and thus should be modeled as well.

Chapter 3
Talker Quality in Interactive Scenarios

In passive scenarios, people just listen (and watch) stimuli, which allows the participants to concentrate well on the task, and facilitates careful preparation and manipulation of the stimuli. In contrast to this, interaction introduces several issues, first of all, it induces verbal flexibility, as the participants should not read text out, but have to produce spontaneous speech for real conversations. For the field of acoustically analyzing Talker Quality, these individual differences between utterances in content and duration require robust acoustic parameters. Even more crucial is the case for experiment that includes pre-defined conditions to be manipulated. In human–human interaction (HHI), many of the features studied in Sect. 2.3 cannot be automatically manipulated in real-time towards the desired target, such as a specific intonation contour, at least not in high quality. In this case, testing different conditions is as difficult in HHI as it is in human–computer interaction (HCI). One solution is to carefully prepare stimuli, and subsequently to constrain the required verbal responses of the interlocutors or computer agents by defining restrictive tasks and scenarios for participants. This, however, does only apply in laboratory procedure avoid of face-to-face situations (see WoZ method in Sect. 3.3). An alternative is to use well-trained confederates for "conversing" with the participants, and verifying the realization of the target conditions post-hoc. However, interaction is of course much richer than the aspects and parameters taken into account in passive listening. All these exciting aspects of the structure and duration of the conversation, of the coordination and feedback by so-called turn-taking and back-channel that involves all interlocutors [94], or the content and expression of intentions of the conversational partner have been absent of controlled in studies presented in Chap. 2.

© Springer Nature Switzerland AG 2020
B. Weiss, *Talker Quality in Human and Machine Interaction*, T-Labs Series
in Telecommunication Services, https://doi.org/10.1007/978-3-030-22769-2_3

3.1 Methods and Instruments for Studying Human Conversation

In order to elicit valid conversation for analysis and modeling, the dominant methods are to either record real conversations of a task-based and well-defined domain in the field, such as counseling interviews or service hotlines, or to invite people into the laboratory and ask them to play such conversations. For the latter, scenarios and tasks are prepared and handed to the participants, who have to mutually solve them. If not only the analysis of the resulting conversations is of interest, but certain hypotheses are to be tested, one of the interlocutors, in the "field" or in the laboratory, could be instructed to behave in a certain way. There are, of course, also non-task situations, where social conversations are recorded and elicited. In order to select comparable material for modeling, the social situation of such non-task conversation should also be defined or restricted. One example is the German Conversation database (GECO), comprising unacquainted females in a laboratory setting, who were instructed to talk about a freely, self-chosen topic for about 25 min [320]. Own contributions, however, are covering task-based laboratory studies only.

Established task-based methods comprise the Map-Task, where two people [9] take over the two roles of the instruction giver and follower, respectively. Both participants are handed a special version of a map, which the other cannot see. The versions were designed to stimulate negotiations, e.g., by varying the amount of and/or by diverging information. There are several conditions to be found in the corpora produced by this task, for example, concerning the ability of the participants to see each other or not. The Map-Task has been used to collect a variety of corpora at different places and in languages. Another method is the Diapix task (a blend of "dialog" and "picture"). It is a dialog elicitation technique that involves pairs of participants cooperatively finding the difference between two pictures; again without seeing the material handed to the interlocutor [22]. In the "Winter survival scenario," the interlocutors take the roles of members of a rescue team. They have to negotiate the priority of a list of 15 items to take with them.[1] Standardized tasks from the field of telecommunication aim at evoking typical situations occurring over telecommunication networks, such as the short conversation tests [163] or the random number verification test [191] (see [163] for example). In the former, typical negotiations of appointments, bookings, etc. are conducted in a dialog with different roles, e.g., a patient calling for setting a doctor's appointment. In the latter, there is no scenario, but lists of random numbers are transmitted to the interlocutor, in order to mimic such special situations that are characterized by a very short utterances and no context for recovery in case of misunderstandings, e.g., due to channel degradation. All these methods bring prepared material to elicit and facilitate dialogs of a certain kind, topic, and to ensure comparability between the sessions. But

---

[1]http://ed.fnal.gov/arise/guides/bio/1-Scientific%20Method/1b-WinterSurvivalExercise.pdf.

there are also tasks for group conversations. In the NASA's "Lost on the moon" scenario [260], for example, a small group of participants has to negotiate a ranking of 15 items in terms of importance for survival, similarly to the dyadic "Winter survival scenario." For the short conversation tests, there also exist versions for three or four participants. As a common characteristic for all these conversational scenarios/tasks, they are solvable, as long as the participants behave collaboratively in the first place. While there might occur sessions with uncompleted tasks, other methods elicit a wider distribution errors or grading task success. For example, the "Lost on the moon" scenario and the random number verification test allow to quantify the success of the conversation, in terms of appropriateness of the resulting list for the NASA task, or by summing up the errors and repairs. But in the end, such methods allow for eliciting authentic conversations in the laboratory. While the NASA scenario expects a face to face interaction, and the short conversation test implies a technical communication service, the other methods can be used either way, as long as the material is not shared (MAP-Task, Diapix, random number verification). In order to assess Talker Quality, which was not the primary aim during development, evaluations have to be included in the procedure, either by external ratings from observers of the recordings [131] or by asking the participants for mutual ratings directly after the conversation [320].

Other procedures are applying behavioral implicit methods from an online ball tossing game and the "Lost on the moon" problem [204]. However, these were used to study the impact of a (passively obtained) first impression, actually the relationship component for explicit and the affective priming task for implicit liking (see Sect. 2.1.1), on the follow-up interactive behavior. So, instead of the Talker Quality formed during the interaction, the interactive result of a passively obtained attitude is studied here. Using a variant of cyberball game [408], the participants are made believe to interact with the known group members they passively experienced before. But in reality, it was a fair computer program. The participants have to decide to whom to play the virtual ball, once they have received it. All players are represented by a photograph, which currently restricts this method to audio-visual Talker Quality. In order to elicit implicit actions, they have the secondary task of remembering a number. In addition, the participants played the NASA scenario, which was recorded, and each dyadic liking was assessed from external evaluators. As a result, the frequency of ball tosses correlates with implicit and explicit liking, while the attributed liking based on the recorded behavior correlates with implicit liking only [204]. For further studies, the procedure of first assessing Talker Quality in a passive scenario could be replaced by a group conversation, followed, e.g., by the ball tossing game.

The interesting aspect of the two approaches, assessing Talker Quality after an interaction or passively before, is the impossibility to entangle the interaction of a quick first, acoustic-(visual) impression that affects interactive behavior of entrainment, turn-taking, and so on, from the recursive impact of such signals in conversation on the Talker Quality, at least with the procedures and methods presented. The conversational features outlined below thus will reflect both, the result of Talker Quality and its origin. It could be argued, however, that for a short conversation, Talker Quality will be relatively stable, given the persistence outline in Sect. 1.3.

Having recordings of conversations by applying one of the methods mentioned above, or similar ones, is just the first step for analyzing and modeling Talker Quality. Another requirement is to extract the features of interest, which includes manual or automatic segmentation and labeling of the recording, or a mixture. Therefore, annotated corpora represent a valuable resource that can be analyzed multiple times with regard to different research questions. In the following, a selection of relevant features that only emerge from the interactive scenario will be presented. These features relate to the structure of the conversation and reflect the communicative style and strategy of the interlocutors. In principle, the features to quantify such styles are also used in HCI (see Sect. 3.3).

3.2   Talker Quality and Conversational Behavior

The communicative styles and strategies used in conversation do affect the interpersonal evaluation and thus the Talker Quality. Such styles can be analyzed from authentic conversations, or can be produced by confederates to increase or decrease the success or even Talker Quality itself [21, 43, 125, 195, 268]. In recent years, a mature level of understanding of conversational processes of dialog partners engaged in spoken interaction has been reached. For the scope of this book, two relevant aspects are the process of prosodically entraining in pitch and speech tempo, as well as the frequency, position, and realization forms of verbal dialog phenomena like back-channel and turn-taking signals.

Linguistic entrainment represents a frequently observed phenomenon of two interlocutors mutually adapting with respect to syntactic, lexical, segmental, or suprasegmental speech production [44]. Most results on it concern the lexical [45, 46] or the syntactic level [297], but rarely, e.g., pronunciation [275] or acoustic-prosodic aspects [225]. Entrainment might be related to the general observation of a positive effect of similarity [212] as described in Sect. 1.3. It is even argued that such entrainment is necessary for mutual understanding [283, 284]. It typically happens quickly, i.e., early in dialog [20]. Despite approaches arguing for an automatic imitation process grounding the entrainment phenomenon [66, 211, 283], e.g., due to priming, its degree is found to vary, e.g., due to social role or interpersonal attraction. This points rather against an automatic process, but not necessarily for a conscious one, if unconscious social moderators exist. For the negotiation of lexical choices in dialog an automatic process is questionable [45]. Even data for "dis-entrainment" is reported [276, 322]. The degree of entrainment seems to reflect aspects of interest and cooperation and relates to task success [77, 211, 298] or smoothness of the conversation [266], and its evaluation [42]. Linguistic entrainment is even observed in humans conversing with machines [44]. An explanation for this relationship between entrainment and evaluation might be the automatic motor congruence facilitating entrainment with a positive attitude or positive relationship [100].

For estimating the effects of the frequency and position of back-channel or of the smoothness of turn-taking on individual dialog success and quality, typically the deviation from a naturally occurring average is postulated. For example, a lag between -100 and $300\,\mathrm{ms}$ during turn changes is considered as barely noticeable, while pauses of the interlocutor a second before responding are regarded as too long and thus might trigger inquiries in the case of an answer expected [94]. The studies presented in the following deliberately assessed the impact of such signals on evaluations, although usually not in such a fine-grained temporal solution.

3.2.1 State of the Art in Talker Quality in Interaction

There is empirical evidence for the degree of phonetic, i.e., typically prosodic, entrainment, as well as realizations of turn-changes and back-channel affecting likability ratings of human interlocutors. So far, these phenomena have been typically addressed in isolation in order to learn about the mechanism involved, and have not focused on evaluation, such as Talker Quality. Other work on interpersonal attitudes incorporating multiple phenomena is sparse and restricted to human interaction [131, 210, 415]. On a conceptual level, proximity, convergence, and synchrony have been separated for studying entrainment in dialog with prosodic cues; finding entrainment dominantly for proximity and synchrony at turn level, with different measures of entraining or aligning at different stages in dialog [223]. Acoustic-prosodic entrainment is further presented here, as it covers primary prosodies such as average F_0, rate, but also intonation contour [130], and thus many articulatory and acoustic aspects considered for Talker Quality in passive scenarios (Chap. 2). But also segmental phonetic similarity has a positive impact on interlocutors' attitudes [124, 289, 334], maybe due to the social relevance of pronunciation for social and regional background and identity.

Interlocutors seem to prefer dialog partners who entrain, e.g., in terms of liking [217, 222, 320, 321, 335]. Interpersonal attitude between couples can be modeled automatically by parameters of acoustic entrainment [217]. However, the processes involved seem to be complex, as prosodic entrainment evolves dynamically over time and even reflects local developments of dis-entrainment [81, 276, 321]. While mostly, a turn-based view on acoustic-prosodic entrainment is applied [223], there are also local mechanisms evident, for example, for pitch in turn-taking [148]. Also, an effect of entrainment is not always evident. For the Map-Task corpus, for example, prosodic measures fail to reach significance in correlation with task success [352]. Temporal entrainment is also evident in overlapped speech [410] and preceding back-channel [129].

However, for back-channel and turn-taking not only entrainment affects Talker Quality, but also the mere amount of such signals or their temporal placement. Likability ratings, collected from observers, correlate positively with filled pauses and contractions, and negatively with, e.g., interjections [131]. Interruptions and, surprisingly, back-channels have a negative effect though. Concerning just the role of a follower, those interlocutors are rated better that take turns by causing overlap instead of a pause. This indicates that overlap might be a simple feature to represent smoothness in turn-taking.

The concept of "cohesiveness" represents positive mutual attitudes. it is loosely related to Talker Quality, and it was studied for the four-person "scenario" data of the multimodal AMI meeting corpus by Lai et al. [210]. Within a number of extracted parameters, averaged post-meeting ratings of cohesiveness are negatively correlated with interruptions, proportion of silence in turn-taking (both in line with [131]) and with the number of the dialog act "eliciting information" (see below). It is positively correlated with turn-taking freedom, i.e., the proportion of each person following a certain speaker, and the dialog acts "providing assessments" and "comments about understanding."

For the Columbia Games corpus, dyadic conversations were elicited by playing collaborate computer games, being located visually apart. Positive back-channel has an effect on the smoothness of turn-changes (low latency in both, pauses and overlap as an alternative feature to the binary overlap), and the game scores obtained during interaction [266].

In [360], turns, silence, laughter, and back-channel are manually annotated for dyadic telephone calls based on the "Winter survival scenario." Talker Quality, in terms of a post-conversational questionnaire of social attractiveness, shows positive correlations of the role of the caller with the number of laughter and back-channel. For the role of the receiver there is a negative correlation with laughter. The number of overlapping speech reveals a similar systematic as laughter, but it is not significant.

Finally, Talker Quality of interlocutors during interviews was compared to third-party observers and related to hand-annotated turn duration and response latency of the interviewee (along with speaking rate) [336]. However, there is no covariation for ratings from participants and observers and participant's data is significantly more positive than for observers. For the interlocutors, response latency correlates with social attraction for both roles. Additionally, interviewees are rated more negatively with increasing turn duration and decreasing similarity in turn duration.

Apart from such descriptive surface features of the conversational structure, a multitude of other aspects characterize a dialog and can subsequently be examined in relation to Talker Quality. For modeling, the quantitative approaches are of more interest than the aforementioned styles that are realized, for example, by confederates. One interesting level of description is the communicative function of an utterance, which drives the conversation, and relates to the intention of utterances, called dialog acts, as reported already above with [210]. In 2012, a standard was published on the definition and multidimensional characteristics of dialog acts [155].

In the referenced study on the AMI corpus, also dialog acts were annotated. Cohesiveness is negatively correlated with the number of the dialog act "eliciting information," and positively with "providing assessments" and "comments about

understanding" [210]. Apart from that study, dialog acts are usually not operational-
ized for studying the effect of conversational strategies/behavior on interlocutors
attitudes, although they are an established level of description and analysis for
spoken HCI (see Sect. 3.3).

In summary, increased entrainment, including the acoustically relevant prosodic
entrainment, but also other aspects of conversational styles affect Talker Quality
and other evaluative concepts positively. There seem to be a discrepancy between
participants' evaluations and those of observers. As there are only few studies asking
participants directly [210, 336, 360], the own contribution on HHI in an interactive
scenario collects such interpersonal ratings. The single negative result of the number
of back-channel was with 3rd-party ratings [131], which is why this parameter is
examined closer in the next section. Finally, overlapped speech was found to affect
positive attitudes [210], which would have a most relevant indication for HCI, as
the prediction of user turn ends, and preparation of system's prompts requires a
new, incremental architecture.

3.2.2 Own Contributions to Talker Quality in Interaction[2]

The aim of this study is to obtain further insight into the strength of correlation
between Talker Quality, directly assessed from the participants themselves, and
the interaction behavior for the domain of unacquainted persons. This behavior
is quantitatively specified by features, such as number of turns, that aggregate a
conversational flow and relate to its structure. They are accordingly called *interaction parameters* [237]. In order to separate the impact of the single conversation
from earlier experiences, the very first impression obtained after the training
session is used as control. Entrainment was not considered here, as currently,
there are only few laboratory systems supporting acoustic-prosodic or segmental
phonetic entrainment (see Sect. 3.4). A laboratory situation was considered to be
most appropriate in order to collect likability ratings and interaction parameters,
respectively, the interlocutors' style, in a controlled manner with regard to situation
and topics, but also to ensure high recording quality for manual annotation. Groups
of three persons were recorded, verbally interacting according to prepared scenarios.
The assumed benefit of triads over dyads is the possibility to compare pair-wise
effects to the likability of a speaker averaged over both interlocutors.

3.2.2.1 Procedure

Altogether, 39 persons took part in this experiment (9 women, 30 men, aged
36.2 years, SD = 12.2). The triads did not know each other in advance. The

---

[2]Parts of this section have been published previously [391].

participants were all experienced in telephone conferences to ensure familiarity with the situation and technology. As a requirement, all had conducted at least three conferences within the last 12 months. On average, the participants stated to have conducted 34.7 telephone conferences during their lifetime. All participants of a group were instructed to the procedure together and then sent to their individual sound proofed room [162]. From this personal meeting, participants could form a first impression of each other, and during the immediately following training session.

The three rooms were connected by a conferencing system implemented in Pure-Data [288]. It provided broadband connection with intensity attenuation of 23.1 dB SPL (test signal of 61.3 dB SPL). Closed headsets of the type Beyerdynamics DT 290 were used by the participants.

Prior to the actual session a first training scenario was conducted. Following this, issues concerning the procedure could be clarified, but the subjects did not leave their individual room. Nine different scenarios were taken from the collection described in [167]: These semi-structured, particularly designed tasks for three-party conversations, similarly to the short conversation tests for dialogs (Sect. 3.1), provide business conversations with topics like choosing a conference location or songs of a music album. Each conversation was initiated by a melody, as the experimental set-up did not allow for a real call initiation.

For each scenario, every participant received different information to contribute or ask for in order to stimulate the conversation. All aspects were "solved" in this manner by the three participants. Although the scenarios provide various job descriptions, no specific (conversational) roles are defined. During annotation, the transcribers gained the impression that roles of leading or following the conversations were taken individually, if at all, and its conversational consequences are thus reflected in the resulting conversational parameters. The training scenario was the same for all groups, whereas the actual scenario accounting for the analysis varied. The conversations analyzed here are only the first block of an experiment with optimal network conditions. The later blocks investigated the effects of transmission delay [316], which is why the scenario was randomized.

After the conversation, each participant was asked to state each partner's likability, as well as personal attention to the call and overall quality of the transmission. The Talker Quality collected after the training is used as basis to control for the first impression established so far. The scale used was continuous with the two antonyms "very likable" and "very unlikable" afterwards transformed to numeric values between zero and ten.

Unfortunately, some participants failed to rate the likability in some cases; maybe due to the cover story of speech quality assessment: For some data, participants did not fill-in the likability scale: From the 39 participants, 3 interlocutors did not rate at all, and 4 of them missed one rating.

3.2.2.2   Data and Annotation of the Surface Structure

The conversations were manually segmented and annotated with ELAN.[3] Two students of linguistics conducted the annotation by supervision of two scientists [391]. The aim was to describe the structure of each recording to quantify for each participant the type of utterance (back-channel, turn, noticeable pause within turn) and turn management (turn change by pause, verbal overlap, or a failed attempt), as found in the related work section. Due to the limited number and limited practical experience of the independent transcribers, no separate annotations with subsequent κ value calculation. Instead, consequent discussions between the two annotators with counter-insurance by the supervisors were decided on to obtain a single confirmed strategy.

From this segmentation and annotation 14 parameters have been extracted to represent structural aspects of the conversation for one talker, denoted as S for speaker in order to not be confounded with turn. Additional parameters can be derived from these, e.g., total time of S turns to all turns, or the ratio of S pause duration to S speaking duration. See Appendix C for the full list of interaction parameters.

Annotating all conversations was very laborious, even though relatively easy labels were used in comparison to, e.g., dialog act labels. Therefore, alternative parameters estimated automatically are used for a replication of the analysis with hand-annotated data in order to test an automatic modeling with such feature estimates. These interaction parameters are extracted based on voice activity detection [234]. Subsequently, segment borders and segment durations have been calculated based on a state model [153]. However, a distinction between back-channels and actual turns is not trivial and was therefore not attempted for this data. Therefore, the automatically obtained surrogate of "a turn" includes all contributions of one interlocutor.

A first inspection of the data reveals potential outliers for the averaged data as well as for the relative ratings, i.e., the difference between the initial rating and the final likability: These are visible as bumps for values over 4.0 (Fig. 3.1), which are actually identical and represent two values. Both outliers differ from the mean more than two standard deviations, and are excluded for further analysis as they might distort the regression models. Interestingly, there is no relevant agreement between two raters on the third interlocutor (intra-class-correlation, $ICC = 0.23$, $p = 0.11$). This holds also for reciprocal ratings ($ICC = -0.25$, $p = 0.91$), which is unexpected, as it is not in line with results for reciprocal interpersonal attraction [15]. As a result, interdependence does not have to be considered in the linear models [185]. According to our expectations, however, there is a moderate correlation for the initial (pre-conversational) and post-conversational likability ($r = 0.54$, $t(50) = 4.58$, $p < 0.0001$).

---

[3]EUDICO Linguistic Annotator: http://tla.mpi.nl/tools/tla-tools/elan/.

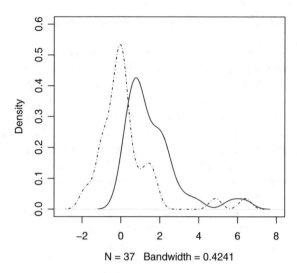

Fig. 3.1 Density distribution of the averaged ratings

The results are presented in two parts: First, the relationship of likability averaged for both conversational partners with surface parameters of the rated person will be examined; second, pair-wise relative parameters for each of the three pairs of interlocutors will be calculated.

3.2.2.3 Averaged Talker Quality and Absolute Parameter Values

Talker Quality of an interlocutor averaged for the two other persons constitutes the dependent variable. Input to the model are the interaction parameters extracted from annotation as well as the initial Talker Quality just minutes before this session, after the training. Data from 35 participants is used, with 7 single ratings instead of averaged ones due to missing ratings.

In a first step, simple correlations are used to include only related variables ($p < 0.10$) in the multiple linear regression to avoid over-fitting during the parameter selection. Thus, from originally 36 variables extracted from annotation, only four items were used in a step-wise multiple linear regression in addition to the initial Talker Quality (number of S turns ending with overlap, ratio of S turns ending with overlap to S speaking duration, ratio of S turn to all turns, ratio of S turns ending by pause to all turns ending by pause). The final regression model includes the initial Talker Quality and two of the four parameters based on AI-criterion ($F_{(3,31)} = 13.66$; $p < 0.0001$, $R^2 = 0.57$, $R^2_{adj} = 0.53$, $RMSE = 0.59$) confer Table 3.1 and Fig. 3.2.

A related model for both interactive parameters only (without the initial Talker Quality) is still significant, but covers considerably less variability ($F_{(2,32)} = 4.09$; $p = 0.026$, $R^2 = 0.20$, $R^2_{adj} = 0.15$, $RMSE = 0.80$). It has to be taken into account that building a model without initial Talker Quality might have led to a

Table 3.1 Results of the multiple linear model for averaged Talker Quality—from annotation

Parameter	Beta	t-value	p-value
Initial Talker Quality	0.56	5.131	<0.0001***
Number of S turns ending with overlap	0.29	2.671	=0.012*
Percentage of S turns to all turns	−0.37	−3.375	=0.002**

Fig. 3.2 Averaged likability ratings representing Talker Quality modeled with initial Talker Quality and two additional parameters

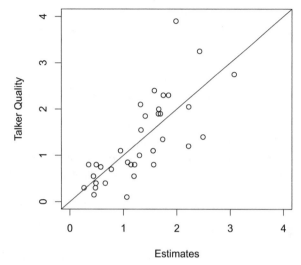

model with more parameters and a better fit, if the task had been to estimate a model based on interaction parameters only.

Most of the explained variance is covered by the initial rating, but parameters describing the surface structure of the conversation also significantly contribute to the final likability. Only two parameters are included in the model, while an increasing proportion of the turn number has a negative effect, the increase of number of turns ending with overlap has a positive one. Interestingly, there is no observed effect for back-channels.

Automatically obtained estimates of the two parameters used do correlate significantly with the respective variable extracted from the manual annotations (Pearson's correlation: number of S turns ending with overlap, $r = 0.60$, $p < 0.001$; percentage of S turns to all turns: $r = 0.75$, $p < 0.0001$). Despite providing only moderate correlations, a repetition of the model estimation with automatically obtained parameters results only in a slightly lower fit than with manual data ($F_{(3,31)} = 11.15$; $p < 0.0001$; $R^2 = 0.52$, $R^2_{adj} = 0.47$, $RMSE = 0.87$), see Table 3.2.

Table 3.2 Results of the multiple linear model for averaged Talker Quality—automatic estimates

Parameter	Beta	t-value	p-value
Initial Talker Quality	0.63	5.278	<0.0001***
Number of S turns ending with overlap	0.33	2.754	=0.010**
Percentage of S turns to all turns	−0.33	−2.812	=0.008**

3.2.2.4 Individual Talker Quality and Relative Parameter Values

In order to test whether the averaging of the likability ratings removes information on the relationship between interaction parameters and Talker Quality, an analysis similar to the previous one was conducted. However, this time individual ratings were chosen as the dependent variable, as well as pair-wise relative parameter values, to cover each of the three pairings of Talker Quality. To represent each pair of interlocutors within one triad, the ratio of absolute values has been calculated as independent variables, e.g., ratio of average turn duration of S to the one of the raters Sx, i.e., S/Sx (see Appendix D). For counts, the number of pair-wise turn events (changes and attempts) is considered for each individual pair, e.g., number of turn changes from S to Sx and vice versa, as well as the ratio of both variables (each value shifted by one, as the data did contain zeros). See Appendix E for this list.

Data from 27 participants is available, resulting in 54 pairs, from which the above mentioned outliers are removed already. This results in about double the data points compared to averaged ratings, as every participant is mostly rated twice, but this of course also induces more variability.

Following the same procedure as for the averaged ratings, simple regressions reveal four parameters out of 14 derived ($p < 0.10$) in addition to the initial Talker Quality: (1) ratio of total turn duration of S to the one of Sx, (2) ratio of turn number of S to the one of Sx, (3) ratio of duration of doubletalk excluding back-channel ($S \xrightarrow{Sx} S / Sx \xrightarrow{S} Sx$) and (4) ratio of number of turn changes with overlap excluding back-channel ($S \to Sx / Sx \to S$).

The subsequent multiple linear regression including the initial Talker Quality and three additional parameters is significant. It provides a fit comparable to the one for the averaged ratings ($F_{(4,47)} = 18.79$; $p < 0.0001$; $R^2 = 0.62$, $R^2_{adj} = 0.58$, $RMSE = 0.83$), cf. Table 3.3 and Fig. 3.3. The respective model without the initial rating is also significant (see Table 3.4), but covers much less variability ($F_{(3,48)} = 5.33$; $p = 0.003$; $R^2 = 0.25$, $R^2_{adj} = 0.20$, $RMSE = 1.16$).

Automatically obtained replacements of the parameters do again correlate with the manually extracted ones (ratio of turn number of S to the one of Sx, $r = 0.83$, $p < 0.0001$; ratio of total turn duration of S to the one of Sx, $r = 0.97$, $p < 0.0001$; ratio of number of turn changes with overlap for both, $r = 0.31$, $p < 0.05$). The subsequent fit is considerable lower, but exhibits a similar structure (see Table 3.4; $F_{(4,47)} = 9.63$; $p < 0.0001$; $R^2 = 0.45$, $R^2_{adj} = 0.40$, $RMSE = 1.35$).

Table 3.3 Results of the multiple linear model for individual Talker Quality—from annotation

Parameter	Beta	t-value	p-value
Initial Talker Quality	0.83	6.346	<0.0001***
Turn number of S/the one of Sx	−0.43	−3.205	=0.002**
Total turn duration of S/the one of Sx	−0.34	−2.652	=0.011*
Ratio of no. of turn changes with overlap	0.41	3.301	=0.002**

Fig. 3.3 Individual likability ratings representing Talker Quality modeled with initial Talker Quality and three additional pair-wise parameters

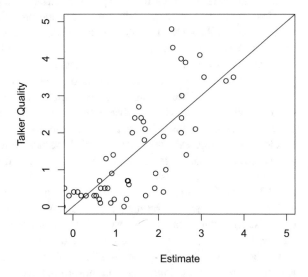

Table 3.4 Results of the multiple linear model for individual likability ratings—automatic estimates

Parameter	Beta	t-value	p-value
Initial Talker Quality	0.80	5.288	<0.0001***
Turn number of S/the one of Sx	−0.28	−1.659	= 0.104
Total turn duration of S/the one of Sx	−0.26	−1.511	= 0.137
Ratio of no. of turn changes with overlap	0.20	1.355	= 0.182

3.2.2.5 Summary and Discussion of the Results for Interaction Parameters

Neither the negative effects of back-channel and interruptions from [131] nor the positive effect of back-channel reported in [360] could be confirmed. Instead, back-channel has not effect, such as in [210] who conclude that very brief assessments are more important than the number of back-channel. However, a positive effect of a turn-transition by overlapping speech [131] is found as well, albeit for the turn holder, not the taker, and in contrast to [210]. It can be interpreted as the participant's degree to invite fluent turn-taking or collaborative behavior. A negative effect of longer turn duration [336] is only reflected in the pair-wise data, not in the averaged data. Instead, an increasing number of turns have a negative impact on averaged and pair-wise ratings alike. Of course, these studies, including our own, are not directly comparable. The way of assessing evaluations and the roles of the talkers were different, which might be the reason for diverging results.

There are two, respectively three parameters significantly contributing to the description models. Although a less conservative approach (i.e., choosing from all 35/14 parameters while ignoring the initial Talker Quality) might lead to more parameters and thus more explained variance, the method used ensures a reasonable number of parameters for interpretation and is adequately fitting the status of the Talker Quality assessment. Also, one major conclusion is that the very first impression should not be neglected in favor of interaction parameters in order to obtain a valid model. Nevertheless, despite the dominating effect of the Talker Quality before the actual conversation, aggregated parameters describing the conversational structure are significantly complementing the description of Talker Quality. This is in line with the observation that the first impression is hard to come by (Sect. 1.2). Using more interaction parameters while ignoring subjective ratings would certainly increase the relevance of those descriptors, but might also lead to a lower fit, as we know, for example, from PARADISE models (see Sect. 3.3) using subjectively described task success versus instrumental measures for the domain of HCI [241]. Still, the aim of this study is to test whether interaction parameters add to initial subjective ratings, thus motivating to study the impact of vocal (and visual) surface features as in Chap. 2 in combination with the conversational structure. For this, there is evidence from the results. It would be interesting to study the relative impact of conversations of the attitude development over time and encounters. The obtained data is, unfortunately, not suitable for this analysis, as the consecutive sessions were highly affected by transmission degradation, for example, asymmetrical delays of several hundreds of milliseconds. These actually not always attributed to the network service, and thus affected the person perception, which renders the data invalid for this aspect, but even more valuable for studying the relevance of communication networks [316].

Both models (averaged and pair-wise ratings) perform similar, with the pair-wise model having one parameter in addition. Both contain a measure for number of turns and one of turn durations. So, speaking less in general is positively regarded in the averaged and individual model. Despite the rich body of results for similarity, this does not support an effect of similarity. This is in line with [210], who also found no effect of "participation equality" in terms of speaking time. The third parameter included the number of overlapped speech during turn change, which might reflect the smoothness of turn-taking.

Indeed, the mere parameters describing surface features of the conversations alone do not provide sufficient information to properly interpret the resulting parameters, especially because replicating studies are lacking so far. Reproducing the results in experiments, e.g., with confederates might be a fruitful approach for future investigations. One important finding, however, is the property of these parameters, to be easily automated and thus estimated with high validity. Just the overlapped speech at turn changes was not estimated satisfyingly, which should be improved in the future. Other interaction parameters might be more difficult to extract automatically. For example, separating back-channel from interruptions or turn-grab attempts would most likely require verbal analysis by automatic speech recognition and language understanding/classification. But, back-channel was not included in the models.

Although the interactive situation is somewhat artificial, goal-oriented conversation could be enabled. Even though, the corpus can certainly not be called representative, it can still be regarded as authentic, which is a prerequisite for studying the relationship of conversational surface structure and Talker Quality.

Several important aspects are not considered in this initial analysis, i.e., aspects of speech acts and politeness, but also group processes and roles. The method itself, namely using triads to assess two instead of one rating can be improved further by collecting acoustic only initial judgments in a passive scenario, and by omitting the training session instead, which does not seem to be required for these experienced participants.

3.2.2.6   Talker Quality and Dialog Acts[4]

Based on the provided literal transcriptions and recordings at hand, manual annotations of dialog act classes were conducted, applying the ISO recommendation [155]. As with the surface annotations, consequent discussions and counter-insurance were conducted to ensure consensus in a single final label. The same two students as before conducted the annotation under supervision. The ISO recommendation explicitly names nine dimensions of dialog acts and specific communicative functions that can either be assigned to a dimension, such as /autoPositive/ for Auto-Feedback, or be general, such as /answer/. The nine dimensions are Task, Auto-Feedback, Allo-Feedback, Turn Management, Time Management, Discourse Structuring, Social Obligations Management, Own Communication Management, and Partner Communication Management. For the specifics of the telephone-conference, a dimension from Annex F.1 of the standard, Contact-Management, was introduced to capture dialog acts of testing and confirming a working line and attention by /contactCheck/ and /contactIndication/.

Dimension 7 was solely annotated for the purpose to test for delay-induced differences [316, Chap. 7], and is not analyzed here. Some dimensions were annotated on two separate tiers to facilitate the work flow. However, during the planning phase of the annotation, not only the dimensions and functions to annotate were defined, but also it was decided to define two new labels in order to cover aspects of interest. The dialog act /failedTurnGrab/ was introduced to cover not-successful interruptions with the goal of taking the floor in addition to successful ones (/turnGrab/). The /turnKeep/ function was not annotated, as it is understood to cover particular pre-planned keeps of the turns, instead of reactions to /turnGrabs/.

---

[4]The work in this section was significantly revised compared to [389] in terms of data preparation and analysis.

The /selfInducedStop/ was defined to cover all instances of unfinished turns, for which the repairing function /retraction/ was not considered appropriate, as this label indicates withdrawing from the own contributions in the same turn without noticeable coordination with or impact from others. This /selfInducedStop/ is actually a subversion of the function /turnRelease/ with an unfinished utterance, such as "…, or …". These kinds of stops are of special interest for the conditions of delay [316, Chap. 8], which are not analyzed here.

Deciding on the extent of annotations and applying the ISO norm was a challenging task indeed. During this process and during the first annotation session, several adjustments were made, and examples collected as references. In particular, the semantic annotations turned out to require much more resources in comparison to annotations of surface structures (such as speaker changes w/o overlapped speech, identifying back-channeling from turn, transcriptions).

Based on the referenced literature in Sect. 3.2.1 and the lacking impact of back-channel in the last analysis, feedback and meta-communication is in focus this time. In order to add annotations of dialog acts to the database, the following dimensions were included as separate ELAN tiers with the specified labels of communicative functions. All labels include identifiers of the speakers to separate the three interlocutors, and in the special cases of Own-Communication-Management and /turnGrab/ also the ID of the addressee, which doubles the number of parameters for these dialog acts:

1. Allo-Feedback

    - /alloPositive/ (indication of assuming the other correct interpretation, e.g., "Exactly.")
    - /alloNegative/ (indication of assuming the other incorrect interpretation, e.g., "You did not get that, no?")

2. Auto-Feedback 1

    - /autoPositive/ (indication of own successful interpretation, e.g., "OK")
    - /autoNegative/ (indication of own non-successful interpretation "Come again?")

3. Auto-Feedback 2

    - /autoSummary/ (rephrasing of an utterance [155, p. 39].)
    - /autoRepetition/ (repeating, "echo" of an utterance [155, p. 39].)

4. Turn Management

    - *NEW:* /failedTurnGrab/ (a /turnKeep/ with attempted /turnGrab/)
    - *NEW:* /selfInducedStop/ (which is ending a turn, but not a /retraction/)
    - /turnGrab/ (talker 2 from talker 1)

5. Own- and Partner-Communication-Management

    - Own-Communication-Management (/selfCorrection/ and /retraction/ from an utterance)

- Partner-Communication-Management (/correctMisspeaking/ and /comple-tion/ of another talker having the floor; talker 2 for talker 1)

6. Contact-Management

- /contactCheck/ (if the other one is listening or still in the line)
- /contactIndication/ (signaling that one is paying attention or still in the line)

From the annotated conferences, the number of occurrence of each dialog act is extracted. In order to limit the number of features for the regression analysis to relevant predictors, some of the parameters are meaningfully grouped, e.g., /selfCorrection/ and /retraction/ summed up to Own-Communication-Management. As most of them occur only infrequently, i.e., fewer than once per talker on average, the individual perspective, i.e., the addressee of a dialog act, was not considered, but only summed occurrences. The final list of eleven predictors include /alloPositive/, /alloNegative/, /autoSummary/, /autoRepetition/, /failedTurn-Grab/, /selfInducedStop/, /turnGrab/, Own-Communication-Management, Partner-Communication-Management, /contactCheck/, and /contactIndication/.

In order to inspect the data, initial linear models were fitted including all 15 dialog act classes. The final data set comprises, including a few missing ratings, 70 data points, in a paired fashion, so that each talker of a triad is related to all two interlocutors concerning the dialog act frequencies. By this data format, no relative frequencies have to be calculated, as at least for Own-Communication-Management and /turnGrab/ there are the partner-directed numbers taken into account. The final linear model with Talker Quality associated with the initial Talker Quality and S number of dialog acts includes only the initial Talker Quality. A model without it includes /autoNegative/, and /contactCheck, but it is not significant ($F(2, 62) = 2.47$; $p < 0.093$), which contrasts the results when using interaction parameters that are representing the surface structure.

This missing relation between dialog acts and Talker Quality is surprising. Maybe, dialog acts do reflect more the attitude of the interlocutors towards a talker, as stated in the beginning of Sect. 3.1. Therefore, in addition to examining the target variance [185], the "perceiver variance" can be modeled: (Is there a relation between someone *rating* positively and his/her own dialog acts?). In the Brunswikian lens models, instead of externalizations, the rater's behavior is taken into account (Fig. 3.4). By taking the observed dialog acts as potential expressions reflecting of Talker Quality that is directed to another talker, using initial Talker Quality is conceptually not grounded. The subsequent model includes six parameters (Table 3.5) and performs moderately well compared to models of perceived Talker Quality with surface features ($F(6, 57) = 5.62$; $p = 0.0001$; $R^2 = 0.37$; $R^2_{adj} = 0.31$; $RMSE = 1.0$).

Of course, including the correlated initial ratings improves the model further ($F(7, 56) = 7.73$; $p < 0.0001$; $R^2 = 0.49$; $R^2_{adj} = 0.43$; $RMSE = 0.90$).

From the dialog acts taken into account, numbers of allo- and auto-feedback have been found to be potential candidates to express Talker Quality. This kind of feedback annotated here was not considered in the previous analysis, where

Table 3.5 Results of the multiple linear model for likability ratings—Dialog Acts of the rater

Parameter	Beta	t-value	p-value
/alloFeedbackPositive/	−0.52	−3.72	<0.001**
/alloFeedbackNegative/	−0.41	−2.90	=0.005**
/autoFeedbackPositive/	0.35	2.40	=0.020*
/autoFeedbackNegative/	0.34	2.32	=0.024*
/autoSum/	0.32	2.22	=0.031*
/selfStop/	−0.43	−1.81	=0.007**

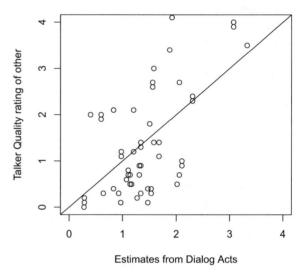

Fig. 3.4 Talker Quality ratings of others modeled with dialog act parameters of rater

only surface information was considered, not separating different kind of feedback. There, only the target ratings were analyzed, for which turn-changes with overlap correlated positively and amount of turns and speaking time negatively, suggesting a positive impact of smooth turn-changes and non-dominant or efficient conversational participation on the raters.

Here, however, positive ratings on the interlocutors are positively related to auto-feedback, including summaries. These kinds of back-channel indicate attention and being directed towards the talker and thus expresses an overall evaluation dimension [252]. A negative impact is observed for allo-feedback and /selfStop/. Negative /alloFeedback/ indicates communication issues that are related to the likability ratings. However, the negative impact of positive allo-feedback is surprising. To speculate, it might reflect the expectation towards the talker to be guided these very easy tasks, in a patronizing way, or it might hint at earlier communication problems. Here, more qualitative analysis is required. Last, occurrences of /selfStops/ represent incomplete sentences. The negative impact might reflect uncertainty on social level, or may hint at non-smooth conversations that increase cognitive load.

3.2.2.7 Conclusion

Just like acoustic parameters representing the voice and speaking style of human talkers, conversational behavior can be quantified and studied on its effect on Talker Quality. Aggregations of rather simple surface features show significant relations with Talker Quality—in our data the number of turns and the speaking duration, even when considering previous ratings. However, no effect was found for back-channel signals. The lack of relevance in such feedback behavior may lie in the rather crude way to quantify it. Apparently, there is a culturally dependent window within the first second of response time that determines whether the turn-taking was smooth or not, and even some overlap will still be perceived as smooth [94]. That means, the current automatic way of separating overlap strictly from pause does exclude smooth turn-taking within the first few hundred milliseconds of overlap. Given the corpora already available, a toolbox for automatic processing and labeling of such parameters that are motivated less technically, but by linguistic and ethnographic research would indeed be very helpful. Such parameters would include, for example, mutual or non-mutual laughing, markers of politeness, and even conversational roles in order to set the parameter values in relation to. Conceptually, the perceptual dimensions and attributions, as defined in the lens model (Sect. 1.2) have to be studied in more detail to provide a proper list of potential interaction parameters. In particular, attributions, such as unwillingness indicated by a longer delay or a particle such as "uhm," are potentially affecting Talker Quality directly.

The attempt to better incorporate functional aspects of conversation by anno-tating dialog acts was not successful in identifying further correlates of Talker Quality. This is in line with another study on multi-party conversation, where also the majority of variance in ratings was explained by surface features [210]. Given the large number of dimensions and potential dialog acts to study, however, the relevance of dialog acts for Talker Quality cannot yet be negated. Interestingly, such interaction parameters did reveal a relation with the quality ratings given by the talker, not the one receiving it. As a result, a model was found for the ratings given by the interlocutors, expressing their attitudes towards the other persons. The number of conversations analyzed is rather low for such a statistical approach, which might cause the low amount of variance explained. That the raters' data shows a relationship with observable behavior is in line with results from [17], who found that extra-version and self-centering of the raters correlates positively with the ratings giving to others. Therefore, it is necessary to further investigate the relationship between dialog act occurrences and personality as well as the expression of individual Talker Quality in terms of relationship variance [185] by dialog acts in the future.

This exploratory study aims at finding candidates from hand-annotated dialog acts correlating with Talker Quality. Interaction parameters provide a compact way to represent conversational structures by aggregating log and annotated data. While the effort varies for the creation of such parameters, multi-party conversations provide an interesting procedure to compare averaged and individual ratings of

Talker Quality. Some of the parameters can be automatically estimated with simple acoustic models with high validity. However, the effort has to be reduced by automation in order to increase the amount of data. Work on automatic dialog act classification has, for example, used word-based features, such as phrases, syntax [197, 218, 296, 340], including the likelihood of a specific dialog act in a sequence by n-grams [333], but also acoustic-prosodic features [323, 413]. There exist also word-based and acoustic models to annotate other complex aspects of utterances, especially those related to humor [292] and sarcasm [56].

Such automation is not only necessary to increase the power of the quantitative models. More important is to take into account the social roles that define a conversation, e.g., caller/initiator/leader from others, as they differ in their acoustic-prosodic styles [324]. Also, the status of observable interaction behavior as either an expression of Talker Quality directed to interlocutors or as cause to others Talker Quality towards the speaker has to be studied in detail. One such exemplary aspect is back-channel, which signals interest, amenability, and attention, all three potentially affecting the evaluation of the person, while also signaling the current attitude towards the speaker having the floor. It would be interesting to identify those interaction parameters that only express Talker Quality, and those that only affect it, if those exist at all.

3.3   Methods and Instruments for Studying Human–Computer Interaction

The approach of annotating or logging human conversation and aggregating the labels by, e.g., summing them is also applied in human–computer interaction (HCI). Here, the established major evaluation concepts are the overall quality attributed to the system and the user satisfaction assessed in retrospection immediately after interaction. While these concepts represent critical aspects of quality of experience (see Sect. 1.1), there is a current development towards addressing more and more the attitude towards the system for speech interaction. The reason is the increasing relevance of the social aspects of the systems for the users, as current systems exhibit more social and human-like characteristics on the surface, e.g., with increasing embodiment, and in their interaction, by pro-active behavior and by addressing personal domains of health care, home services, or other private domains (see Sect. 1.4).

3.3.1   State of the Art in Methods and Instruments in HCI

Aggregated data on conversations are called interaction parameters and have been extensively used to model user satisfaction in a linear fashion for the domain of spoken dialog systems. In the 1990s and the millennium, the dominant application domain was telephone-based self-service. This domain establishes highly task-based HCI situations. For such task-based and thus goal-based interactions, enabling a successful conversation can be easily defined by any kind of well-defined task completion, such as a bank transaction and a positive database response. The systems during that time were geared to the traditional definition of usability:

Usability: The extent to which a product can be used by specified users to achieve specified goals with effectiveness, efficiency and satisfaction in a specified context of use. [156]

In other words, usability is "getting the job done," with minimal resources (in time), and by taking into account individuality of the users and their current situation. User satisfaction is assumed to result from a successful and efficient conversation. Consequently, an instrumental model to describe and predict user satisfaction was defined, the PARADISE model, for which user satisfaction is considered as task success minus "costs" [369].

As the concept of Talker Quality is typically not assessed in studies on task-based spoken dialog systems, despite early work on social actorship [295], user satisfaction is considered as the most related concept for presenting this part of the state of the art. In PARADISE, task success represents effectiveness, whereas the costs comprise measures of efficiency. Interaction parameters can be used to quantify both parts, although task success is sometimes defined by user ratings just like the user satisfaction itself, or by expert's ratings. Some of these efficiency-related interaction parameters are separated from "efficiency measures" by being called "quality measures" [369], as they seem to have a direct relation to perceived quality, such as (annoying) response delay and the amount of specific (e.g., repair) actions. While this distinction is not very precise, it raises awareness of potential differences in impact of the interaction parameters that go beyond overall dialog duration, and already hint at potentially relevant perceptual dimensions.

An instrumental measure of task success is, of course, just an estimate that may not always represent user perception very well. Consequently, applying instrumental measures results in a performance drop of the models compared to user-reported task success [237]. For the case of form-filling dialog systems, where the dialog is driven to collect a number of required and pre-defined attribute-values for a database search [170], kappa (κ) is an established value of task success (see Eq. (3.1)).

$$\kappa = \frac{P(A) - P(E)}{1 - P(E)} \tag{3.1}$$

From a confusion matrix of intended and received values for stored attributes of a form, $P(A)$ is the probability of correctly filled attributes, and $P(E)$ the chance level. By this equation, κ denotes task success corrected for chance level, and its calculation is based purely on the recognition of pre-defined attributes. In the case of an unknown distribution of chance, it is advised to estimate $P(E)$ from real data [369]. From user experiments in the two domains of telephone-based self-service and smart-home, PARADISE models reveal to be very delicate, as they cannot be used to predict user satisfaction when changing the user group, system configuration, and the domain (confer Sect. 3.4).

A recommended list of interaction parameters defined and used in various studies with spoken dialog systems is available from the International Telecommunication Union (ITU) [165]. The parameters can be separated into five classes:

1. Dialog- and communication-related parameters

   - durations, such as user turn duration
   - quantities, such as number of words and utterances
   - flow of information, e.g., number of questions or number of new concepts introduced

2. Meta-communication-related parameters

   - communication problem, e.g., help requests, system no-match prompts
   - recovery, such as correction rate

3. Cooperativity-related parameters

   - number of in-/appropriate system prompts according to Grice's maxims (see below)

4. Task-related parameters

   - user ratings of task success
   - experts decision on task success
   - κ, as presented above

5. Speech-input-related parameters

   - speech recognition, e.g., word and sentence error rate
   - speech understanding, such as number of correctly filled attributes of a form.

The third class covers parameters that are addressing cooperativity of the system. Parameters from Class 3 are representing, for example, the outcome of badly designed prompts and dialog flows. This approach is based on work of Grice [134] who defines principles of cooperativity that are assumed by interlocutors in order to produce and interpret contributions to a dialog. One example would be "be relevant," which allows interpreting replies in HHI that are not directly related to a question as a meaningful answer nevertheless. Take for example [78]:

A: Is there another pint of milk?
B: I'm going to the supermarket in five minutes.

Only under the assumption of relevance, the implicit reply of B is considered as appropriate and the correct pragmatic meaning can be derived. For the domain of spoken HCI, these maxims are extended and guidelines are derived to facilitate cooperative system behavior [34, 90]. By checking for "violations" of these guidelines, cooperativity can be quantified. However, in HCI these principles are used to create explicit utterances to avoid any potential misunderstanding, even the milk example. This contrasts HHI and results in unnatural system utterances with the aim of maximum functionality. Of course, in HHI, underspecification and context-relation is an intrinsic behavior, which increases efficiency. Therefore, in HHI the example above must not be considered as violation, but as a perfect example of the power of Grice's maxims. One of the reasons for this discrepancy is manifested in one of the new principles provided for HCI itself, the partner asymmetry that is characterized by the imbalance of capabilities and knowledge shared between human and computer. It requires the system to be explicit about its capabilities, at least as long as such human-like implications and indirect utterances are not possible to be properly processed by computers.

Not only cooperativity, but also many other parameters do require manual annotation, such as recognition related parameters. This holds also for dialog acts (Sect. 3.2.2.6), which have been applied to HCI, for example, for the cases of exchanging the required attribute values of a task-based system, such as requesting, providing, confirming, and acknowledging key variables (e.g., destination of a time travel information service) [366]. Using this kind of interaction parameters can increase performance of PARADISE models [368]. These dialog acts are also the typical abstract concepts for representing meaning and actions in spoken dialog systems [73, 170]. As human-like turn-taking and back-channel is still not available for speech products, parameters related to these, which have been analyzed in human interaction (Sect. 3.2.2.2), are not included in this list.

In order to systematically study interaction parameters and their impact on evaluative concepts, such as user satisfaction or Talker Quality, the Wizard of Oz technique (WoZ) can be applied. With this method, one or several system modules are replaced by a human operator (the wizard) unknowingly to the users of a, typically laboratory, system. While WoZ is also important during the development process to collect data for machine learning modules or to test design choices without having all modules ready, it also allows for efficiently testing system configurations systematically. For example, a human can try to achieve perfect language understanding. By artificially introducing errors, parameters from Class 5 can be systematically tested on their impact on users.

Closely related to instruments for evaluating HCI are perceptual dimensions of and attributions towards conversational systems, as both represent relevant stages of information processing, association, and, in the case of attributions, even evaluation-based classification (refer to the lens model in Sect. 1.2) that may affect Talker Quality. Ideally, such dimensions and attributions are represented in comprehensive questionnaires that are not only aiming for overall Talker Quality, but its individual components, as presented for HHI in Chap. 2. For task-based spoken dialog systems, seven "quality aspects" have been separated in a review of analyses of several empirical studies applying various questionnaires [236]. These aspects comprise attributions according to the lens model, e.g., personality or cooperativity, but also aspects that are not related to the system, but the interaction and even the user themselves:

- Acceptability: How successful, functional, and difficult the interaction was; this is strongly related to user satisfaction and overall quality
- Communication efficiency: The dialog progress in terms of time
- Cognitive effort: The cognitive resources required from the user
- Personality: Attributions of politeness or intelligence
- Smoothness: The ease of use, but also perceived control over the dialog
- Cooperativity: The attributed teamwork as outlines with Grice's maxims
- User factors: Such user characteristics determining if the specific user fits to target user groups, the system was developed for, e.g., language proficiency.

While all these factors potentially affect Talker Quality, acceptability—by its characteristics of preference, interaction pleasantness, and future usage—comes closest to it as single overall attitude. All these quality aspects are assessed in the comprehensive framework recommended by the ITU [164]. Questions regarding the system's voice and personality are, however, restricted to naturalness and politeness, or indirectly referred to by asking for the enjoyment and pleasantness. In sum, this questionnaire is a detailed, but long, and rather functional-analytical instrument suitable for diagnostic research of task-based applications. Much less functional are questionnaires for anthropomorphic systems, where emphasis is given on social aspects, such as social presence, human-likeness, and liking (Sect. 1.4). For the case of robots, for example, the "Goodspeed" questionnaire [25] assesses *Anthropomorphism, Animacy, Likability, Perceived Intelligence*, and *Perceived Safety* on bipolar, 5-level scales. Perceived Safety is clearly special to autonomously moving robots and do not apply to fixed interfaces, such as embodied conversational agents (ECAs). For a review of relevant empirical methods and evaluation concepts and their applicability for ECAs, please confer [402].

3.3.2 Own Contributions to Methods and Instruments in HCI[5]

A questionnaire that bridges the gap between the functional view on spoken HCI and the social view was developed for ECAs [379]. It is based on the aspects *satisfaction, engagement, helpfulness, naturalness and believability, trustworthiness, perceived task difficulty, likability*, and *entertainment* [305]. As high-level aspects, satisfaction and engagement comprise other aspects, for example, trust and likability in the case of engagement. Conceptually, these meta-aspects can be seen as resulting from the attributions towards the ECA and assessment of the interaction, although likability should belong to them as well, at it represents the resulting overall attitude towards the artificial talker. In two experiments with German participants, a new questionnaire was developed and validated that addresses these aspects, except for satisfaction and engagement, which should be derived from the basic aspects. Two

---

[5]This section is a summary of the four publications [166, 175, 379, 402].

different facial and voice models were used in four conditions (see Sect. 2.3.3) in a WoZ laboratory experiment that required to fulfill seven tasks related to a smart-home system [404].

The dialog flow was controlled in the following fashion: The tasks were written on separate cards and offered to the participants in a pre-defined order. Each participant had to carry out both scenarios once with each version of the ECA. To avoid boredom the tasks were slightly altered in expression and content while the level of difficulty of each task remained constant. The order of scenarios was varied between participants. After each scenario, a brief quality assessment was conducted. After both scenarios, participants had to fill in the questionnaire based on the complete interaction with one version. After revising the initial questionnaire with 51 items based on the results for 50 participants, a second experiment was conducted with 24 participants using only one combination of a voice and, this time, a stationary, but physically embodied head [378]. This validation confirms the sufficient reliability of the sub-scales in terms of Cronbach's α: Likability $= 0.84$, Entertainment $= 0.74$, Helpfulness $= 0.94$, Naturalness $= 0.82$, Trustworthiness $= 0.82$, Perceived Task Difficulty $= 0.80$. The resulting German version of the questionnaire sufficiently assesses the six quality aspects, with the Likability/Talker Quality scale comprised of the items on pleasantness, likability, friendliness, and agreeableness. A version for Australian English was validated with a setup similar to the first experiment, but with 18 participants, just changing the voices to English. This English version, however, should be modified according to the changes proposed after analysis [402], particularly to increase reliability for trustworthiness (from $\alpha = 61$ to 0.76) and Entertainment (from $\alpha = 68$ to 0.79). In any case, the new version should be tested again in order to ensure the positive effects of the revision.

Most of these aspects, in particular trustworthiness, naturalness, and entertainment are not intuitively linked to aggregated dialog structures in form of interaction parameters. For these attributions, there may be other features than interaction parameters that represent aesthetics, personality, and so on. For voices and speaking styles, such acoustic parameters have been studied in passive scenarios (Chap. 2) and they are not in scope in this chapter. Nevertheless, there are some cross-correlations between the factors of the questionnaire [379], and also, aspects of the dialog are expected to correlate with Talker Quality based on results from HHI, such as number of turns and dialog durations (Sect. 3.2.2.2). This is the topic of the following sections.

However, while there is an extensive list of defined interaction parameters for task-based spoken HCI, some modifications and additional extensions are necessary in order to address specific multimodal aspects of HCI. For a proper description of the interactive sessions, the supported modalities have to be treated separately, and not only on a multimodal, abstract level, e.g., separating verbal from gestural interaction. Therefore, it should be supported to relate error rates and numbers of system and user information produced to each modality. Apart from this separation of some interaction parameters for each modality, there are eight additional parameters [166], three of them adopted from other research on multimodal HCI:

1. *System Feedback Delay* (SFD) is the delay from the end of a user's action (utterance, gesture, ...) until a perceivable reaction from the system. SFD allows to separate the first perceivable reaction from the beginning of a full response in terms of executing a command or answering a question by delivering content expected by the user. SFD differs from the established system response delay (SRD) [165] for signals of understanding and processing, such as "OK, ..." or "Uhm, ..." for voice, an hourglass mouse pointer or progress bar for displays, or a smartphone vibration during typing on a soft-keyboard. It can be assessed instrumentally.

2. *Modality Appropriateness* (MA) [33] captures how suitable the offered modalities are for a specific input from a specific user in a specific situation, and accordingly the same for the system output. In order to decide on appropriateness, modality properties [33] or guidelines from multimedia learning [243] can be consulted. Appropriateness can be labeled binary, on a scale, of, as recommended [166], as fully, partially, or not appropriate. Manual work is required either for labeling or defining system-dependent instruments for automatic assessment.

3. *Relative Modality Efficiency* (RME) [279] is defined as number of information bits, for example, attributes in a form filling system, that are correctly transmitted in a given duration (e.g., turn or time) separately per modality. Individual modality efficiency can be related to each other for estimating relative efficiency, or to its relative usage for assessing the appropriateness of the modality selection or user interface design. This parameter can be implemented for instrumental assessment.

4. *Multimodal Synergy* (MS) [279] is the proportion of the time to complete a task in a multimodal interaction to the expected time without any synergy. The latter duration is estimated from single modality session durations, which are weighted according to their relative usage in multimodality. A non-synergistic reference is required for calculating the relative part of it.

5. *Number of User Modality Changes* (UMC) reflects the frequency of modality changes. It can be separated for user (input) and system (output) changes. This parameter could be automatically extracted from log-data.

6. *Lag of Time* (LT): Related multimodal system output that express jointly the same intention/information has to be synchronized in time, just like facial expression, gestures and utterance of humans are temporally planned and produced in synchrony [193]. Given a simultaneous or strictly sequential presentation, the lag of time measures the divergence from the intended presentation as overlap or delay. In addition to appropriate log data, a planner module is required that explicitly states the intended temporal relation between the system output modalities.

7. *Number of Asynchronous Events* (AE) is derived from the Lag of Time and sums up the number of non-optimal multimodal presentations. This parameter requires context dependent thresholds of appropriate lags of time.
8. *Number of Unsupported Modality Usage* (UMU) counts the number of user input attempts that cannot be processed by the system for this modality or modality combination.

An initially proposed *Fusion Accuracy* is the accuracy of the fusion module only. Hence, it reflects the final accuracy that is relevant for the systems' response, while separate accuracies would report on the individual or aggregated performance of the individual recognition modules. It is defined in [175], but it was not included in the recommendation [166], as it is similar to Understanding Accuracy (UA) [165] if UA is calculated after the fusion module, and not separately for each input classifier. All these new interaction parameters are directly related to quality by their definitions, except for UMC. While definitely an important descriptive parameter, UMC might either reflect the optimal usage of the system modalities provided, or it might indicate recognition problems. Therefore, this interaction parameter can be considered to be highly prone to system changes or even user (group) differences with respect to Talker Quality.

How meaningful these parameters are for a given speech-based human–computer interface depends on the modalities supported. For a purely anthropomorphic interface that aims a natural face-to-face interaction, Talker Quality is not to be expected to relate to modality appropriateness, which should be rather constant for a single system. There are only few instances where developers and interaction designers really have to choose the appropriate modality, e.g., speech or pointing for signaling spatial information. For systems that provide additional interfaces, however, such as the pepper robot does with an integrated screen that even allows touch gestures, modality appropriateness will be much more relevant. This extension of interaction parameters from spoken to multimodal interaction is a step towards standardizing definitions, annotation schemes, and parameters for multimodal HCI, which previously have been distributed over various larger projects [35, 122, 231, 332, 346].

3.4 Talker Quality in Interaction Between Humans and Computers

3.4.1 State of the Art in Talker Quality in HCI

Just like the rich resources on entrainment in HHI (Sect. 3.2.1), the phenomenon of users entraining is also found in interaction with dialog systems. Most work have been done for lexical or syntactic entrainment [44, 72, 230], resulting in specific computational models [46, 59, 131, 286].

Students' prosodic entrainment to tutoring systems in pitch [347] and in pitch and intensity [373] correlates to the users' learning gain. Speech rate adaption of users is found also in spoken interaction with embodied systems [28, 272]. As the presence of these phenomena has been shown to be positive in HHI, there has been some activity in building systems with such capabilities. However, the degree of pitch entrainment was lower for synthesized speech compared to pre-recorded natural speakers [347], raising the question whether an adaptive system will have a positive effect at all.

Actual results on the effect of adapting systems are rather sparse so far: Implementing a speech rate adaption algorithm based on users' speech rate and response time was preferred over a system with constant rate for Japanese [374]. Testing a repair strategy for problematic user speaking styles, such as over-articulating and shouting, one system version was built that exhibited adaptive behavior. It reduced intensity or increased speech rate, which did successfully induce users' entrainment and thus promoted a normal speaking style, leading to better speech recognition performance [99]. Entraining a tutoring system in pitch increased naturalness and rapport perceived by observers [232, 233]. An implementation of entrainment on multiple prosodic cues (intensity and rate) in one out of two helping advisors in a card game resulted in higher explicit liking, and in higher implicit trust measures (i.e., the number of asking this helper for advice) for the entraining system [222]. This latter system has been recently expanded to Spanish and Slovak [224]. Still, the small number of systems and experiments can only be a starting point in studying the effects of entraining systems on Talker Quality.

The apparent focus in spoken and multimodal dialog systems on user satisfaction and system quality, instead of Talker Quality (see Sect. 3.3.1), is only different for anthropomorphic systems [402, Sects. 1.4 and 1.5]. Even for the high number of interaction parameters (see Sect. 3.3.1) it is unclear, which of these actually contribute to Talker Quality. In the separation of pragmatic qualities and hedonic qualities, it has been argued that pragmatic qualities, comprising usability, and being certainly reflected by many interaction parameters in task-based domains [237] are mere "hygiene factors" that remove barriers. However, in order to actually evoke positive experiences, hedonic qualities are necessary [143]. Apparently, some interaction parameters do reflect hedonic qualities, at least in HHI, such as number of back-channel. Of course, interaction parameters can also deliberately include hedonic signals, such as affect display [314, p. 76 ff.]. For dialogs from the field, not laboratory ones, task success, the number of recognition errors, and the number of system requests explain about 92% of the variance of system quality ratings [97]. However, the evaluation concept are expert ratings of recorded dialogs, not satisfaction scores of the real users.

Models of user satisfaction for applications, such as train timetable or e-mail reader include user's report on task success and the mean speech recognition score, along with others, such as dialog duration (negative), and number of barge-ins or help requests (both negative) [370]. Such models explain between 39% and 0.56% of the data variance [367]. Related results are found in [237], who reports that user satisfaction can be modeled to about 40–50%, with the most relevant parameters being

- measures of communication efficiency, e.g., words per system turn, system turn duration (negative), system response delay (indicating longer user utterances), but not dialog duration
- measures of appropriateness of system utterances, especially the number of appropriate system utterances
- measures of meta communication, especially the number of user utterances that are partially correctly understood, i.e., at least one user attribute, but not all; and number of corrected misunderstandings
- initiative: the number of system questions.

In addition, the number of user barge-in (negative) and the word accuracy contribute to the models. Task success is only included as subjective user rating.

When exchanging the subjective task success with an external or instrumental measure, such as κ, however, the performance drops [237, 241]. There are limits in generalizing such models to other domains, system configurations, and user groups, resulting in rather low R^2 of about 0.20 up to 0.50 [95, 137, 367]. Applying non-linear approaches instead of the linear regression does not increase generalizability [241]. Only in the case of considering user reported task success instead of instrumental measures or external ratings, the performance exceed 0.50 (just below 0.60 [237, 367]). In real systems, collecting user satisfaction and users' view on task success is difficult, though, and will lead to a bias towards successful interactions [97]. Even in laboratory data, expert driven "interaction quality" is better modeled by automatically derived interaction parameters than user satisfaction [314]. The unweighted average recall (UAR) of the support-vector-machine is $UAR = 58.9\%$ for the expert ratings, but only 49.2% for user satisfaction. For the latter, the most important parameters are the number of turns, those related to the recognition performance, user turn duration, number of time-outs, and percentage of system re-prompts. The Spearman correlation between expert and user ratings of task success is reported with $\rho = 0.66$.

The PARADISE approach has also been used for evaluating human–robot inter-action. In a collaborative setting on assembling wooden toys, out of a small set of interaction parameters, the most important ones for various questionnaire categories related to user satisfaction are the number of user repetition requests (negative), dialog duration, and the recall of system instructions (hand-annotated observations of following the provided task guidelines) [117]. The positive impact of dialog duration is argued to stem from the novelty of the robot or the collaborative task. Overall, however, the explained variance is very low ($R^2 \leq 0.20$), so that other factors have to be considered for a proper modeling. For a robot-bartender, Talker Quality was assessed with the Goodspeed questionnaire, but only before the interaction and after the final trial, showing no significant change from expectation to experience [182]. In the experiment, however, a hand-coded and a data-trained policy for verbal and gestural system actions was compared. Its evaluation was conducted on eight interaction parameters, as well as a short questionnaire that was not Goodspeed, unfortunately. For the trained policy, results show a shorter dialog duration only. All questions were answered more positively for the trained system version, for example, questions five (on naturalness) and six (happiness about the interaction).

Other, nonverbal system behavior, have been repeatedly shown to affect the attitude towards embodied systems. While such behavior can be quantified in principle as interaction parameters, it is typically implemented as binary system characteristic and thus considered as experimental condition. There is much empirical evidence of synthesized nonverbal behavior affecting user ratings, among them scales of Talker Quality, such as likability, in a positive way, and evoking increased nonverbal signals on the user's side as well [198].

The visible behavior, however, does not have to be a direct consequence of the attitude in every case, despite such results (see also Sect. 1.4). While Talker Quality assessed on a questionnaire with items on, e.g., likability, warmth, friendliness, and pleasantness is not affected by the smiling behavior of an ECA, the smiling duration of the users increased significantly, when the ECA was smiling [200]. To the contrary, an experiment with a robot displaying head movement and gestures resulted in higher Talker Quality, but in less nonverbal signals [202]. Hence, this kind of nonverbal entrainment, or mimicry, cannot be taken as direct signal of positive Talker Quality. There might be a lacking correspondence between behavior on the one side and perception and evaluation on the other, based even on "sublime" attributions of sociality, that may trigger automatic or learned schemes in users [198, pp. 201 ff.]. In her comprehensive book, Krämer concludes that appearance has an important impact on likability [198, pp. 206 ff.], as it does in humans [254]. However, while smiling can express positive and friendly attitudes, it can act as negative social signal, and it might interact with, or at least not compensate for, other attributions (cf. Uncanny Valley effect, Sect. 1.4). For nonverbal signals in ECAs, for example, to coordinate turn-taking [235] or smile, gaze and proxemics [62], many effects on attributions and user attitudes observed in HHI can be confirmed. Nevertheless, some exceptions presented here indicate limits in applying all results to HCI, which might be caused by the naturalness of their synthesis or even by influences from unconsidered other factors. Just as with HHI, appearance, and voice as well (Chap. 2), represent factors affecting quick attitude formation. Therefore, not only the expansion of interaction parameters from simplistic efficiency-related surface features, such as overall dialog duration, to semantically or even pragmatically founded categories, such as dialog acts, is well advised in order to cover social and cooperative behavior and non-strictly task-based situations. In addition, the first impression formed by such quickly evaluated factors, may they be purely aesthetic, or affect Talker Quality moderated by person attributions, is necessary to incorporate for building more powerful models.

3.4.2   Own Contribution to Talker Quality in HCI

Own HCI experiments comprise two speech-based interactions with an ECA, and two more experiments focusing on state of the art input devices, both for the domain of a smart living room. All four experiments applied WoZ methods in order to replace the natural language processing, and in one case, a camera-based free hand gesture recognition.

In the experiments with an ECA, only speech was provided as user-input to interacting with the agent. The settings were in one case a laboratory test with a simulated smart-home (46 participants, aged 20–60 years, M = 28.9, SD = 7.65) [404] and a realistic experiment in a fully equipped smart living room (49 participants, aged 20–61, M = 29.1, SD = 8.34) [173]. Two different facial and voice modules were tested in four conditions and compared to the passive evaluation scenario (Sect. 2.3.3). The results concern the degree of interactivity, i.e., from passive over simulated to a real interaction in the laboratory living room. With increasing interactivity, the performance of Talker Quality models, using separate quality aspects from ratings scales as input (speech quality, visual quality, synchrony, fit), decreases in terms of R^2 of about 0.65 over 0.50 to about 0.25. A similar effect is observed for the influence of an additional medium, a second screen in the case the simulated smart-home [404]. Both factors, interactivity and additional media, reflect a diversion from the ECA by the interaction, which can explain the performance decrease in the ECA Talker Quality model. In addition to this, Kühnel concludes from her analysis of these data sets that the ratings reflecting Talker Quality are confounded with overall system quality (including the media service, etc.), indicated by correlations of about $r = 0.6$–0.7 [172, p. 72ff.]. A more reasonable argument is the ECA representing the whole system due to its embodied user interface and human-like voice, which are relevant aesthetic factors and invite to be attributed to service aspects, such as recognition performance and dialog appropriateness. This actually seems to be meant by Kühnel's usage of "confound" (p. 76) and is in line with other results from the smart-home domain [240]. As a result, interaction parameters can, and should, be applied to model Talker Quality of the ECA in particular, and not just the overall quality of the whole interactive system and smart home service, as conducted in [172].

For both ECA experiments, a set of interaction parameters was extracted from log data: *dialog duration* (DD), *number of user turns* (#user turns), *system turn and user turn duration* (STD, UTD), *user* and *system response delay* (URD, SRD), *number of user help requests* (#help), *number of no-input* (#noInput), and *no-match* (#noMatch), *number of system questions* (#system questions), *number of barge-in* (#bargeIn), and deviations from an ideal path of the tasks (#leftPath), e.g., by reordering sub-tasks. Task success was not analyzed, as this WoZ scenario was always successful.

For the first experiment, a reanalysis of the Talker Quality ratings obtained from the likability factors of an early version of the ECA questionnaire indeed results in significant models. All parameters have been normalized to more easily compare their respective effects. The first model, using only interaction parameters (Eq. (3.2)), is very poor ($R^2 = 0.06$, $p = 0.012$), but still shows an impact of the interaction on dedicated Talker Quality of the ECA, and not only on overall system quality. A step-wise approach includes user ratings of visual quality (VQ) as subjectively perceived module characteristic (Eq. (3.3)), that has a stronger impact than the interaction parameters ($R^2 = 0.39$, $p < 0.001$), further supporting Kühnel's claim. The negative impact of *dialog duration* and the positive one of *visual quality* is obvious, whereas the positive impact of #leftPath is interpreted as positively regarding the ECA's (actually the wizard's) capability to behave context appropriate, in light of being a surrogate to *contextual appropriateness*, which would have been more laborious and thus was not annotated.

$$QoE = -0.004 - 0.21 \cdot DD + 0.22 \cdot \#leftPath \tag{3.2}$$

$$QoE = 0.006 - 0.26 \cdot DD + 0.11 \cdot \#leftPath + 0.59 \cdot VQ \tag{3.3}$$

A similar reanalysis of the second ECA interaction in the realistic smart living room, also testing the two voice and facial modules, results as well in a rather low performing ($R^2 = 0.12$, $p = 0.020$) model (Eq. (3.4)). This model is rather different in terms of predictors included. There is a negative impact of the *number of system questions*, indicating understanding issues or inefficient user input, a repetition of the positive relation of leaving from the task path, a negative impact of barge-in, where users did not want to wait for the system prompt to finish, but interrupt, a positive sign for #user turns, as well as a negative inclusion of user response delay, indicating processing issues due to user uncertainty. Only the surrogate of *contextual appropriateness* is included in all models.

$$QoE = 0.09 - 0.31 \cdot \#system\ questions + 0.11 \cdot \#leftPath + 0.27 \cdot \#user\ turns$$
$$- 0.18 \cdot \#bargeIn - 0.14 \cdot \#user\ words - 0.15 \cdot URD \tag{3.4}$$

The sign of these parameters is not surprising, except for two. The *number of user turns* is contrasting an efficiency-related impact, and just might indicate that users used some freedom provided to interact more than necessary with the system, which was allowed in the very last task. In this case, it would not represent an efficiency-based interaction parameter, but instead an expression of an already established positive attitude or a positive mood, which is related to positive evaluations [381]. The negative sign for number of user words also needs a comment here, as is may be just representing efficient interaction. However, given known differences between user groups of "factual" users compared to "social users" [411] in interaction behavior and also in ratings of spoken dialog systems [119], this parameter might also hint at a negative bias of "factual" users who produce shorter and less natural utterances.

$$QoE = 0.08 - 0.30 \cdot \textit{#system questions} + 0.11 \cdot \textit{#leftPath} + 0.31 \cdot \textit{#user turns}$$

$$- 0.17 \cdot \textit{#bargeIn} - 0.33 \cdot \textit{#user words} - 0.16 \cdot URD$$

$$+ 0.24 \cdot \textit{#politeness} - 0.14 \cdot \textit{#laughter} - 0.13 \cdot \textit{command style} \quad (3.5)$$

Therefore, this data set was enhanced by additional parameters of social signals, such as laughter, command style, and politeness, related to aspects presented at the end of Sect. 3.4.1 on smiling. The performance increases just a little ($R^2 = 0.21$, $p = 0.001$). As with the first experiment, the same parameters are kept when including additional parameters (Eq. (3.5)). In addition to these, there is a positive sign for politeness markers (German polite version of "you"), a negative of laughter, and also a negative for the binary distinction of having a command style. As with the number of user word, there is a negative relation between command style and the ECA's Talker Quality in the second model. The model performances for Talker Quality of the ECAs are much lower than reported for subjective overall system quality of spoken dialog systems (Sect. 3.4.1), as well as for the same experiments [172]. This indicates at more influencing factors that currently considered with interaction parameters despite the assumed association of the ECA with the overall system and service. Models of Talker Quality of the ECAs show relevant impact of traditional interaction parameter, so not only features that affect quick impression formation (see Chap. 2), e.g., aesthetics, play a role for. Still, as with HHI, such first impressions that can be obtained in passive scenarios, here the quality of the visual model, have a stronger impact than parameterized interaction. This, of course, is in parts caused by inducing variability with different experimental conditions by exchanging ECA modules, which amplifies the impact of such factors. However, also surface parameters aggregating social signals, as politeness markers, improve such a model. This might be factors representing differences in system configuration, or effects of user group differences. Before presenting own contribution on user group, the development of the smart-home system used in the last experiment to offer multimodal input conducted allows for studying the relevance of multimodal interaction parameters defined in Sect. 3.3.2.

In order to address these newly defined interaction parameters, multimodal systems were set up that allow for multiple input devices, such as speech, touch (on a screen), and three-dimensional gestures, in a sequential way. This means that users have the freedom to (mostly) choose with each action the most convenient modality. By this usage of the term, modality is used according to Oviatt and Cohen [271], who state the following on their first page:

> Multimodal interfaces support user input and processing of two or more modalities—such as speech, pen, touch and multi-touch, gestures, gaze, and virtual keyboard.

In two additional experiments, both WoZ, such systems were evaluated for the smart-home domain in a laboratory living room. They enabled users to operate devices in the living room, such as lights, blinds, TV, an electronic program guide, just like the ECA system did only with speech.

There was no ECA used as interface, as the main system output was shifted towards a big screen that displays video/TV content, as well as for displaying system information and navigation. Instead of an obsolete ECA questionnaire, other questionnaires were applied, which aim mostly at usability and are therefore not presented here. An exception to this is the AttrakDiff questionnaire for assessing quality of experience [144] that was applied to assess retrospective evaluations directly after completing scenario-based sets of tasks. The overall attractiveness of the interactive system is assessed along with the sub-aspects of pragmatic and hedonic quality, the latter divided into identity (the fit of the system to one's self-image) and stimulation (how exciting the interaction is). Overall attractiveness comprises the items "good–bad," and "ugly–attractive" in a short version and in addition the scales "likable–unlikable," "pleasant–unpleasant," "repudiative–inviting," "encouraging–disencouraging," and "repelling–appealing" [87] (own translations), which does not directly address Talker Quality in terms of an attitude towards the system, but closely relate the quality of experience that is correlated to Talker Quality by its definition (Sect. 1.1) and empirically, for example, for ECAs ($r = 0.86$, $p < 0.01$) [379]. This questionnaire was chosen because of its wide-spread use in the scientific community as alternative to instruments for Talker Quality, as the interactive system provided multiple user interfaces, speech being just one among them. The underlying assumption of attractiveness being a linear combination of the two hedonic qualities and the pragmatic quality has been sufficiently confirmed in each own data set collected with this questionnaire in the past, further validating this instrument.

In order to apply the new set of multimodal interaction parameters (Sect. 3.3.2) two additional experiments were conducted with the system providing multiple ways of user input. QoE is modeled from normalized data of these two interaction experiments. In manual annotation and semi-automatic processing of log data from the various devices, six of the eight new parameters are combined with a subset of extracted interaction parameters for spoken dialog systems. The new ones comprise MS, RME, UMC, UMU, LT, and AE, but not SFD and MA due to restricting in the already significant amount of data preparation. In both experiments, users were asked to operate the smart-home system first with all provided input modalities one after the other in separate sessions, before conducting the multimodal session that is used for modeling. For details on the annotation, please confer [172].

In the first experiment [174], speech and a smartphone user interfaces were provided, allowing for 2-dimensional gestures on touchscreen and 3-dimensional gestures by moving the whole smart-phone. Twenty-seven participants (aged 21–35, $M = 26.1$, $SD = 4.05$, 10 of which are males) took part. A step-wise inclusion into a linear model results in three parameters. These are *number of speech user turns* (#user turns$_{speech}$), *incorrectly parsed utterances* (PA:IC), and *dialog duration* (DD). The resulting linear model (Eq. (3.6)) reaches decent performance compared to typical values for spoken dialog systems, but outperforms the cited human–robots interaction ($R^2 = 0.52$, $p = 0.003$). For the smart home system, there is a positive impact of using speech, consistent with the efficiency of this modality for operating devices and particularly selecting content from the electronic program guide and media server. The negative effect of duration and incorrect processing of user input is also related to the efficiency experienced with the system, and is, most likely, originated in the task-focused situation of this laboratory experiment.

$$QoE = 0.06 + 0.40 \cdot \#user\ turns_{voice} - 0.43 \cdot PA{:}IC - 0.29 \cdot DD \qquad (3.6)$$

The second experiment was conducted with a group of 17 younger (20–29, $M = 24.6$, $SD = 2.74$) and 16 elder adults (51–67, $M = 58.8$, $SD = 4.45$), gender balanced [176]. Applying the same parameters as with the first experiment, no significant model is obtained. However, a similar analysis results in three different parameters included, *task success (TS)*, *number of user turns with touch* (#user turns$_{touch}$), and *error rate gestures* (ER$_{gesture}$) (Eq. (3.7)). Task success was not included previously due to its lacking relevance. This experiment, however, showed to be more challenging, which could be an effect of including an older user group. The selection of modality related parameters for touch (positively) and gestures (a negative impact of error rate) can be considered as system specific. Their relevance might change with the recognition performance and their specific usefulness to operate differently designed systems. The model's performance is poorer compared to the previous experiment ($R^2 = 0.42$, $p = 0.002$). This, however, is not solely caused by the increase of heterogeneity of the participants, as normalization separately for age group increases the performance only slightly ($R^2 = 0.44$, $p = 0.001$).

$$QoE = 0.03 + 0.38 \cdot TS + 0.39 \cdot \#user\ turns_{touch} - 0.38 \cdot ER_{gesture} \qquad (3.7)$$

There are two remarkable aspects to be mentioned for these two experiments. Firstly, high-level hand-annotated interaction parameters, like those related to appropriateness and especially those dedicated to multimodality, are not included in the models apart from PA:IC. Instead, measures reflecting single modalities are included. This may change for multimodal systems that provide a more sophisticated fusion of user input than the current sequential multimodality. Secondly, even for the same domain, and only marginally different systems, there is no generalizability of the models. Still, the relevance of age group is minimal.

Data from the second experiment was also analyzed in a larger approach to create a screening instrument for a HCI user taxonomy that better reflects user group differences than age or gender. This approach focuses on information and communication technology and uses six, not independent, dimensions (ICT anxiety, privacy related skepticism, ICT competence, orientation on surface features, exploratory learning, ICT interest) to cluster attitudes and stances related to ICT, as well as their individual interaction and overall acceptance behavior [150, 263, 264]. Based on this approach, data from altogether 139 people was analyzed in a confirmatory factor analysis, and further validated internally with self-reported data on usage and ownership of ICT products and services [400]. Apart from this important step towards a German screening questionnaire for assigning participants to ICT-related user groups, this information can support modeling. Interaction parameters from the second multimodal experiment show meaningful significant differences between user groups: Playful and experienced users uttered more help requests (to learn about the system) than all other represented classes. They also interacted longer with the system, when just taking into account the last, more open sub-task of a session, which fits the highest values in exploratory learning of this class. Especially for efficiency related metrics, such a knowledge-based approach to user groups could provide helpful to model user behavior.

So far, all variability in interaction parameters presented stems either from the individuality of the users, and their subsequent differences in interaction behavior, or from basic changes and variability in the system, e.g., the used facial and speech ECA modules or performance changes of the classifiers. But with recent developments in machine learning, user interfaces increasingly adapt to specific users and contexts by explicitly building modules that are supposed to change their behavior over time, which refers to the passive input-mode (see the beginning of Chap. 1). This is especially relevant for ECAs, which aim at mimicking human, and thus user and context adaptive, behavior [168]. One example already mentioned in the frame of verbal interaction is applying context appropriate assumptions, using underspecified references, and even implicit responses (see Sect. 3.3). There might be positive effects due to more appropriate system behavior, e.g., perceived as "intelligence" or "smoothness." However, there might also be negative attitudes emerging by automatic system adaptations, due to a several reasons.

- *loss of control* from the user's side, which can result in cognitive reactance in human social situations [93]
- *confusion* due to an unstable, i.e., unpredictable behavior
- *privacy concerns*, e.g., when a system apparently tracks your behavior
- *inappropriate behavior*, simply due to a wrong generalization of data or an error in classification (most prominent is the example of suggesting a product after purchase of a related one in the case of non-consumables).

For the application of an embodied visitor information systems, such adaptations were tested for their impact on Talker Quality. The system was placed in a showroom to inform about research activities, the institute, and to promote demonstrators and project posters that are located in the showroom. This time, no WoZ was applied. Instead, a working system, including a speech recognizer (push to talk) and implemented adaptations, was tested against a basic system without automatic adaptations, in two subsequent interaction experiments (16 and 30 participants). The first version of the adaptive system exhibits four kinds of adaptations,

- *User recognition* with face recognition, which allows for greeting the users after a break, when they left the ECA to test a demonstrator or read a poster, and asking about the project they visited;
- *Topic preference* by remembering the first choice of project topics, when proposing to inform about an additional project;

- *Amount of information* implemented as binary choice of preferring only one paragraph/prompt on each project or both available; and
- *Suggestions* preferred by the user in contrast to decide on the projects by selecting from a list of keywords.

While no quality model is attempted, the interaction parameters illustrate no differences in speech recognition performance, or in number of projects asked for. However, interactions with the adaptive version are shorter (DD), and apparently invites for a higher *#bargeIn*. Thus, in both evaluations [392, 401], the adaptive version shows significant improvement in efficiency. In contrast to this, the scales used for assessing Talker Quality do not show a significant difference in either experiment, just a positive tendency towards the adaptive system. One reason might be a missing salience of the adaptations, even in the within-subject design applied. Only 7 of the 16 participants could report after completion about the general nature of the differences [392]. A closer look in the second experiment reveals that all participants noticed the user recognition, and all except one the system noticed topic preferences, while the amount of information was not only noticed by 10 of the 30 [401]. The most prominent feature is thus one with the least impact on interaction parameters, but signals contextual appropriateness. The results also exemplify that not all interaction parameters, even in direct comparison, relate to Talker Quality. Thus, more research is required in order to explain and model the impact of system changes and interaction parameters on Talker Quality, especially for the growing amount of adaptations to be expected with the current innovation in user- and context-aware (speech-enabled) HCI. The results presented here indicate an issue of perceptual salience of the system changes, which would mean that the corresponding level to explain the lacking impact would be the perceptual dimensions.

In this book, expectations and stereotypes are only mentioned as potential influencing factors, and studied actively only to define HCI user groups. For speech-based multimodal HCI, for example, mid-aged users are very pessimistic in their expectations on the quality of such a system, but evaluate it similarly to the other age groups in direct retrospection of the experienced interaction [169]. While some expectations stem from real past experiences, others merely have the status of guessing, especially in the case of new technology. Therefore, the first impression has been considered in this book instead of an expectation, not only for HHI (Sects. 1.3 and 3.2.2), but also for HCI. The question for HCI is in this last contribution of this chapter, whether first impressions are valid, or at least consistent, with retrospective attitudes towards a speech-based system, directly after an interactive session.

The first impression was assessed from participants in four experiments, two of them already mentioned as adaptive ECA [401], and the second smart living room system with three kinds of user input modalities [176]. The two new experiments are both with smart TVs as interactive systems, one offering TTS as output, while the other uses only visual reaction on the TV and the smartphone screen. These two experiments were conducted in the frame of applied research projects in order to evaluate speech-interfaces for cooperate products, both using a smartphone as touch and speech interface. The participants were recruited according to the target group (40 participants, aged 20–70 years, M = 38.9, 23 female; and 19 participants, aged 23–50 years, M = 29, SD = 6.3, 12 female) [399]. The first impression was obtained after presenting different system output to each user, i.e., several screenshots (3–7) of typical displays, comprising the main smartphone graphical user interface, the TV/Screen of typical menus and search result views, presented in a random slideshow for a second for each picture. A second is more than sufficient for stimulating consistent first impressions, e.g., as even attractiveness for websites is highly consistent after 500 ms exposure compared to 10 s [351]. For the audio and speech output, selected prompts were recorded and played back with its normal duration. For the ECA system, this was a video instead of pure audio. While such research for GUIs and websites found consistency comparing different systems [4, 214, 227, 228, 351, 353], the four systems here were analyzed regarding within-subject consistency. The first impression was assessed on the AttrakDiff for each output modus separately (smartphone, graphical content on the main screen/TV, TTS or the ECA), just as the quality of experience after the interaction. Results show significant correlations for the overall attractiveness (QoE) ranging from $r = 0.38$ to 0.75. A simple remembering of questionnaire items can be eliminated as reason for the correlations, so that there is evidence for within-subject persistence of the first impression of QoE. In addition, the speech synthesis shows stronger correlations than the two other output devices (smartphone, TV) for the Smart TV system with TTS output ($r = 0.50$ compared to two times $r = 0.36$). However, a comparable experiment with a smart living room, TTS correlates least ($r = 0.38$ compared to TV $r = 0.50$ and smartphone $r = 71$). This unclear order most likely reflects the attention and thus salience of the respective output device for the interactive QoE. For the systems offering not only speech but also other user input, there were also interactive sessions conducted for each modality provided, e.g., GUI and three-hand gestures, before the target multimodal session. When describing the multimodal QoE ratings with linear models of the first impression and the ratings of the uni-modal interactions, both kinds of predictors are included. This suggests some kind of complementary relation of the two sources of predictors. In addition, the AttrakDiff's pragmatic quality is revealing a much lesser or insignificant consistency compared to overall attractiveness. As this does not only hold for the first, passively obtained impression, but also for the uni-modal interactive sessions, it is concluded that pragmatic quality reflects a concept with consistency for the studied systems in general, instead of being a low valid concept just for passive first impressions.

3.5   Conclusion and Future Directions

Interaction parameters aim at representing individual conversations in HHI and HCI alike. While they can be obtained from manual annotations, automatic procedures are handy, especially based on logdata that is available in HCI. Interaction parameters can be separated in categories that are based on the kind or level of information they represent:

1. *surface structure* covers durations and numbers of events that are purely descriptive and thus rely on logdata alone or only on simple models and estimates, such as dialog duration, number and duration of double-talk, and speaking time from a model of speech activity. The number of word and related parameters can also be assigned to this category, if automatic speech recognition for the given language and recording provides high accuracy.

2. *conversational phenomena* are related to symbolic social interaction and thus require interpretation, as they are dependent on human perceptual dimensions (confer Brunswikian lens model, Sect. 1.2). Instead of parameterizing double talk and speaking duration on acoustic models of speech activity as with the surface structure, a model of turn-taking has to be applied to count, for example, turn changes that have to be separated from back-channel, or to count the number of head nods or smiles.

3. *evaluations and intentions* also require models of human cognition. Such parameters address the interpretation of social functions, most prominently dialog acts, as well as attributions and evaluations, such as the number of "smooth" turn-taking and appropriate responses, or even higher-level evaluations like cooperative vs. uncooperative styles.

The results in this chapter reveal the limits for generalizing relationships and effects of such efficiency-related low-level descriptors, rendering quality models useless for changes in domain, user group, and other contextual factors, as efficiency does not apply uniformly, e.g., by diverging references, for changes in these factors. The challenge is to disband such surface structures and replace this category by more appropriate parameters from the remaining two categories. As a consequence, "interaction parameters" should include data on categories and social signals that are typically studied and modeled in passive scenarios, overcoming the separation of social signals and aesthetic aspects presented in Chap. 2 from their occurrences and usage in interaction in this chapter. Only with sufficient data, however, more general models can be build, which means that the effort of hand annotation should always be invested with the aim to create and improve automatic models.

For the case of Talker Quality, there is much less data available from interactive scenarios compared to passive ones, especially concerning perceptual dimensions and attributions relevant for Talker Quality. Even for speech-based interactive systems, functional aspects still dominate evaluations, while the attitude towards the system, i.e., Talker Quality, represents only an established evaluation concept for embodied conversational agents and social robots. Consequently, modeling non-linearities in Talker Quality is currently out of reach. With more data, this will hopefully change and will allow, for example, dimension reduction in order to identify perceptual dimensions. "There exists no comprehensive set of quality aspects for multimodal HCI" [306] expresses a similar conclusion that is still valid today. Methodologically, defining logging and annotation schemes for conversations, along with data aggregation, has to be pursued particularly for multimodal HCI, as done with the definition in Sect. 3.4.2, for developing and improving unified models. While such schemes exist for HHI, there are limits in transferring them to the HCI. One major limit is embodiment. For the multimodal input provided by the systems of Sect. 3.4.2, there are sequential modality changes and pre-defined, i.e., unnatural, categories and commands for input gestures that extend natural interaction between humans and with ECAs. This aspect addresses the essential difference between HHI and HCI, a pre-defined asymmetry between human and computer. On the one hand, systems are still inferior to process social and contextual information, on the other hand, they have access to unique data, as they force humans to apply pre-defined commands and can access, for example, GPS location, external databases, and huge amounts of logging data. A systematic agreement on how to incorporate this asymmetry in quality models is still missing, although it affects the reference of an attitude. One important development in the definition of interaction parameters will affect aggregation functions, which are currently averages and sums, because context and time-dependent weightings are not yet considered in relation to Talker Quality. But the position of social signals, e.g., might strongly affect their impact on retrospective ratings, as is does for QoE of communication channels [406]. For now, results exemplify that adaptivity improving the efficiency of an HCI session may not be relevant for talker quality of an ECA. Such system functions can still be studied in small-scale experiments and in the field. For speech-specific adaptations, for example, linguistic and prosodic entrainment, new results on Talker Quality are expected anytime due to recent developments enabling real-time adaption functionality.

Chapter 4
Talker Quality in Design and Evaluation of Speech-Based Interactive Systems

Scientific theory and practice in HCI has changed over the last decades, not only due to new insights and empirical progress, but also with the emergence of new trends affecting viewpoints and raising new research questions. Well-known systematizations of this development point out major phases and key changes in the scientific community [18, 141, 302]: Starting here with the beginning of 1980s, the first focus was on usability, aiming at identifying and subsequently removing specific obstacles in user interfaces when trying to successfully and efficiently complete a task. Laboratory experiments based on methods and theories from cognitive psychology were conducted for this purpose and affected design criteria. Typical topics would be optimization in selecting and placing icons so that recognition, identification, and usage would be fast and error-free. For speech interaction, similar questions regard the number of options and database results provided to the user and the hierarchy depths of navigation menus. At the end of that decade, the focus was widened from the human cognitive processes to domain/task-specific interactions. This enabled to address more explicitly information representation on various levels involved, i.e., studying aspects of transparency and clarity when providing system and interface information to the users. The claim was, for example, to not restrict design and development decisions to results from laboratory experiments (e.g., limitations in working memory processing), but to include contextual information on the specific situation and the task, which is to be completed, supporting the identification of usability issues, e.g., originating in the integration the work in daily routines. With this turn to understand HCI as social phenomenon, context was becoming more relevant, and new, especially ethnographic, methods and theories were adopted. This change came, of course, also with technological changes of graphical user interfaces and applications of computer-supported collaborative work.

Contemporary developments include, according to [302], four turns: "to design," "to culture," "to the wild," and "to embodiment." All these four turns emphasize individual experiences of users, may this be by providing aesthetic and exiting interfaces, taking into account user group expectations and interpretations, evaluating the

© Springer Nature Switzerland AG 2020
B. Weiss, *Talker Quality in Human and Machine Interaction*, T-Labs Series
in Telecommunication Services, https://doi.org/10.1007/978-3-030-22769-2_4

in-cooperation of modern technology into daily life and studying need fulfillment or life changing effects for the domains of health or sustainability, or reflecting more on the action part of HCI by recognizing the physical manifestation of user and interface in the surrounding world. While the turn to design is, among others, about the responsibility of the creators for their product and service, resulting in discourses on and approaches for inclusion and sustainability, the turn to culture critically reflects on the range of interpretation that is set by cultural contexts of the users and developers. For conversational agents, such critical perspectives concern in particular the attribution level of the lens model, for, e.g., selecting voices and speaking styles: How does the current choice of predominately female voices in in-car navigation systems, personal assistants, and smart-home devices stabilize gender stereotypes and hierarchies? Therefore, awareness is required of which kind of talker attributes are selected, as the resulting service is not only used, but integrated in personal life and thus potentially affecting the range of interpretation.

More practical aspects are to be drawn from the work comprised as "turn to the wild" and "to embodiment." The desire to conduct research in the field and to develop with feedback from the field is fueled by the need to cover complex contextual factors influencing the users' experience. For efficiency-centered cognitive theories and technological advances, the required validity might be reached in the laboratory. However, the target concepts such as Talker Quality, user experience, and quality of experience, authentic contextual situations are, by definition, part of the evaluation concept, which is why the first impression for which passive scenarios have been applied have to be studied in conjunction with interactive scenarios for obtaining a realistic perspective on retrospective Talker Quality after conversation. As a result, testing for effects of a single change (experimental condition) is not necessarily possible with many influencing factors, which are uncontrolled for each user. For analyzing field results, therefore, the identified factors relevant for Talker Quality provide a good basis. But it has to be kept in mind that situated conversations may increase the probability to use the peripheral route, i.e., to apply heuristics and stereotypes and favor the first surface features perceived (see Sect. 1.2).

All these turns also reflect technological advances and evaluation concepts: Nowadays, basic functional qualities can usually be guaranteed when considering best-practice from the last decades, so the subjective perspective of a positive experience makes up the service. With the emergence of natural synthesizers for human voices, faces, and movements, and conversational systems entering public and private spaces, on a smart phone, but also in new devices or embodied in a robot, these four turns also apply to speech and conversational interfaces. A topic rather specific to this increased naturalness of synthetic talkers is the resulting perception as social actors (Sect. 1.4). It is not congruent with the "turn to embodiment" or "to the wild," and is even called "social turn" [365]. For more details on this and methodological consequences, see below (Sect. 4.3).

As a consequence, design and evaluation has to be oriented—not necessarily at absolutely human-likeness—at human-like social and communicative features and behavior. The latest impressive example of synthesizing almost natural voice and conversation behavior relies on domain-specific training.[1] Therefore, the concept of Talker Quality is becoming more and more important in HCI, especially since speech and embodied user interfaces aim at enabling intuitive interaction. For speech in HCI, this means focusing less on disturbances, degradations, or disruptions in speech service provided, favoring instead those aspects of quality of experience that are currently and increasingly receiving more attention. For example, naturalness and aspects related to intelligence, personality, emotion, and sociality are increasingly used for evaluating TTS and interactive systems, which has been the motivation to define Talker Quality in the first place. As the Uncanny Valley illustrates (see Sect. 1.4), artifacts in the production of artificial speech and conversation are not directly constituting the Talker Quality, but moderate references considered during quality formation. In this light, Talker Quality and related concepts are apparently affected by even selected aspects of human-likeness and single social signals, without the requirement of simulating the same level of human similarity for all those characteristics. As several results in this book indicate, many factors affecting Talker Quality in HHI can therefore be in principle transferred to HCI, even in selection. Furthermore, while end-to-end deep learning of agents' conversational behavior can simulate humans amazingly well, it is limited to the data selected for training and therefore difficult to modify. As a developer or a designer, however, being able to systematically select and change aspects (personality, voice) and conversational strategies is a desired feature. It can be argued therefore that a knowledge-based synthesis of specific personality and social characteristics that can be developed and tuned in an engineering or design life cycle represents a meaningful alternative or extension to straightforward end-to-end training. While the trend in industry might move to major global companies, who own sufficient data, computational power, and expertise to providing such services even for small and medium-sized enterprises (SME), the current challenges for SMEs are to create conversational services and interfaces without such support. Minimalist smart-home skills or smartphone apps force developers to rely on the technology provided by the respective platform, but still, there are applications that can consider lessons learned from scientific results as presented in this book.

Results presented in the last two chapters support the claim that both, existing attitudes and interaction behavior affect retrospective Talker Quality. In own contributions these existing attitudes are studied as first impressions that can be collected in passive scenarios and thus cover those impressions that will be established immediately during the beginning of an interaction between unacquainted interlocutors and agents, i.e., during the first seconds. While the present studies have only approached the complex relationship between this very first impressions and interaction behavior in first encounters, results indicate nonetheless that such immediately formed impressions are persistent and influence human (inter-)actions, the latter at least for HHI, according to own contributions. In line with this, results indicate for HCI that just analyzing interaction parameters for modeling Talker

---

[1] https://ai.googleblog.com/2018/05/duplex-ai-system-for-natural-conversation.html.

Quality—even those which might act as mediators of first impressions—is currently not sufficient. Consequently, design and development procedures have to address aspects of voice and speaking style for those that can be evaluated passively, as well as specific conversational aspects. By defining new metrics and applying new classifiers, aspects of first impressions might be assessed in models based on inter-action parameters due to the immanent relationship (Sect. 3.5). However, neither for automatic models nor for non-automatic assessment methods refraining from such an efficient approach of applying passive scenarios and building subsequent prediction models would be meaningful.

4.1   Perceptual Dimensions

Starting with major results that bear relevance for the design and development of speech-based systems, the perceptual dimensions of voices and speaking styles provide a vocabulary of relevant terms, when it comes to describe, separate, and select voice talents for recordings, but also for more technical applications. With the perception of darkness (related to pitch level and spectral centroid), variability in intonation, vocal softness, perception of fluency (even in the absence of clear breaks), pronunciation precision, and tempo, perceptual categories are identified that can be used to characterize talkers, in part even automatically. In particular, softness and intonation are relevant for Talker Quality of female talkers, and softness and darkness for males. Until the availability of robust automatic models, especially for softness, individual experts or even groups of non-expert listeners can be recruited, for determining such characteristics, by applying, for example, the questionnaire in Table A.1. As variability in the fundamental frequency is the most important feature for a large database of longer, conversational, speech samples, the related dimension of intonation (or even the associated person attribution of being active or calm) is particularly relevant for conversational interfaces with more than current one-shot interaction, and of course reading services, such as audiobooks, relaying on speech synthesis.

Technologically, the results and models are also applicable for speech synthesis: As increasingly synthesis frameworks rely on end-to-end learning, the current results and models could be used for proper selection of speakers and data, as mentioned above. Even more interesting would be to test, whether such automatic models could be applied as instrumental evaluation stage in an iterative development process for refining and optimizing such deep synthesis model architectures and parameters. Furthermore, as currently deep learning allows creating almost naturally sounding utterances, changing talker characteristics is an important and active field for adapting a high-quality voice to other identities, personalities, or moods. As such models are usually trained with a lot of data, such modifications should require only minimal additional information, for example, by transforming the value space of the acoustic representation [414] or by applying transfer learning approaches. In order to obtain or retain high Talker Quality, the acoustic parameters found to affect Talker

Quality can be used to decide on the best selection of additional material, or even for determining target values for changing a voice in its characteristics with minimal data.

For conversational scenarios, there does not exist a validated list of perceptual dimensions yet. However, contributions of interlocutors in terms of (relative) number and duration are relevant in both, HHI and HCI. With the current absence of human-like turn-taking, it might be feasible to test the relevance of smoothness in turn-taking, which is even evident as critical event in the simple and automatic measure of double-talk duration for HHI, also for HCI.

There are approaches ready and research systems available that provide the necessary incremental architecture to support human-like turn-taking [26, 326]. The major requirement is to process user speech input already before the user utterance is finished, as it is currently done. While this is technologically already possible, an architecture that allows to use the current hypothesis of the automatic speech recognition module for formulating potential system actions and utterances to play (a) at a smooth temporal distance to the expected user utterance end, and (b) start the system output with only parts being already formulated and synthesized is not yet used in real products. Similar characteristics of an architecture are required to enable an agent to entrain acoustic-prosodically, especially at the temporal positions near turn-taking. For this feature, the perceptual dimension(s) are still unclear, but first studies already indicate a positive effect on Talker Quality. Such architectures can be realized already with existing frameworks and models, not only to test whether turn-taking, backchannel, or entrainment are perceptual dimensions in conversation, but also to verify their impact on Talker Quality.

In Fig. 4.1, there is one illustrative example depicted. The upper part in green illustrates the modules required for a basic incremental system. Please notice the bi-directional communication between natural-language-generation (NLG) and both, text-to-speech synthesis (TTS) and the dialog manager (DM). The TTS module

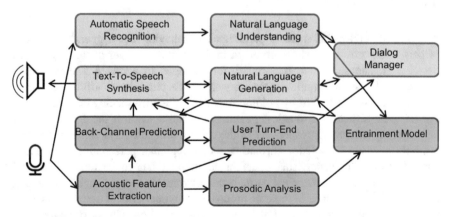

Fig. 4.1 Example architecture of an incremental system providing smoother turn-taking and entrainment

has to give feedback on the current state of production and playback, so that DM and NLU can react by complementing and correcting system output at the correct moments with new information from the recognition side, or when the system output is finished before the complete system action is produced, e.g., by inserting stalling phrases or even hesitations. In the lower part (blue) an additional acoustic analysis is inserted that is not aiming at phonemes or words, but at paralinguistic and intonational aspects for predicting user turn ends, and parameters of acoustic-prosodic entrainment (in the prosodic analysis). The NLG is here taking over the function to produce and feed the surface form of the back-channel, which could be based on the pragmatic and lexical context, to the back-channel prediction module. This and the user end-turn prediction trigger the playback of the appropriate utterances. As turn-relevant moments can be in principle used for either turn-taking and for back-channel, both modules have to know which one has priority, there is a bi-directional information flow connecting them. Finally, the entrainment model receives lexical and syntactic input from the NLU, and acoustic-prosodic data from the special prosodic analysis module, in order to adapt the systems verbal (NLG) or acoustic (TTS) entrainment. A permanent or at least, frequent update of the respective module output, along with rejections of modules estimates and hypotheses that allow the system to work below the time frame of utterances, is assumed.

Entrainment reflects the capability of covering one important aspect of reciprocity, which is also relevant for perceptual dimension, along with turn-taking and backchannel strategy, i.e., the perception of similarity in such dimensions. Interestingly, reciprocity is not only a potential perceptual dimension, but could also result in person attributions, which are represented in the next layer of the revised lens model (Sect. 1.2), if it would be evaluated in friendliness or agreeableness.

4.2 Attributions

Any kind of conversational and paralinguistic behavior that triggers attributions of personality are definitely represented in the attribution layer of the model (see Sect. 2.3.2), and subject to reciprocity, for example, in agreeableness. For conducting well-designed studies, reciprocity should therefore be controlled, either by analyzing it (Sect. 2.1.1) or by holding it constant due to screening of participants. For HCI, the difficulty is to assess potential attribution towards the agent that may be subject to reciprocity, especially in the case of embodied systems, which offer not only acoustic stimuli for attributing social, regional, personality, or affective characteristics, but also visual ones. In addition, it has to be ensured in HCI evaluations that all experimental participants interpret the social situation similarly, and not split, for example, in separate groups of analytical-critical users versus those who put themselves in the position of an everyday situation. This is, of course, true for all evaluations, but in particular relevant for controlling reciprocity effects.

But even more straightforward attributions of talkers' traits and states, such as maturity, interest, and emotions, are associated with positive and negative attitudes (see Sect. 2.2.2). For speech-based material from interactive scenarios, warmth, attractiveness, confidence, compliance, and maturity have been identified, and the first two could be related to Talker Quality (see Sect. 2.3.2). Despite the recording being conducted in interactive scenarios, the ratings were collected passively listening though. Therefore, the acoustic and perceptual characteristics leading to an impression of, e.g., warmth in conversation might be subject to contextual and sub-cultural differences. Also, stereotypical associations with, and preferences of person traits and states are subject to the perspective of the raters, for example, their cultural and regional background, and not studied in this book. The self-evident guideline for this complex topic is to fit the selection of states and traits to be appropriate to the service and target groups. It must be noticed, however, that attributions based on a perceptual dimension might differ in passive and interactive scenarios due to the origin of the signal used for inference. While the passive scenario can be applied to mimic the first seconds of conversation, perceptual dimensions might be evaluated differently in the social and pragmatic context of interaction. For example, synthesized hesitations, useful for grabbing users' attention or keeping the turn while receiving information, especially in incremental systems, do not affect Talker Quality in conversation as assessed with the Goodspeed questionnaire [36], while a negative impact is to be expected without such context, given the results in Sect. 2.3.2.

Attributions towards talkers that are based on conversational behavior are admittedly lacking a structured research agenda. Especially for designing computer agents to interact with a specific communicative strategy, there are typically only singular or a few parameters manipulated, such as impolite verbal instructions for arrogance (Sect. 1.4) or proximity for obtrusiveness (Sect. 1.5). Actually assigning conversational strategies to personalities, stances, or affective states, however, need further research, especially for multi-cued strategies. This includes the representation level of dialog acts, which is used for PARADISE-like modeling, but is lacking evidence to affect Talker Quality even in HHI (Sect. 3.2.2.5).

4.3 Talker Quality

Evaluation and modeling of Talker Quality in conversation has been conducted with interactive scenarios in Chap. 3. While the issue of different categories of interaction parameters, as well as their limitation in considering temporal and positional effects on Talker Quality has been addressed in Sect. 3.5, another, more global, issue has not yet been mentioned: The lack of unified ways to automatically log data. With the "turn to the wild" researchers have to deal with data from the field, for example, from mobile applications. As a consequence of the potentially massive increase in authentic usage data, hand-annotation becomes impossible, even with many crowd-based label and ratings methods. In addition, the increase in mobility for smartphones or even individual placement of smart-home devices results in unknown environments and social contexts. Simultaneously, laboratory-like audio or video recordings of the users for manual annotation often do not exist. In sum, there is a dependency on automatic logging, and thus the need for finding suitable automatic replacements of hand-annotated concepts (Sect. 3.1), or even for building models to estimate those, always considering their ability to replace manually obtained parameters in their relation to Talker Quality.

With this methodological expansion from the laboratory experiments applied here to near or full field tests and experiments, the method to assess Talker Quality also has to change. While own contributions in this book show a significant impact of Talker Quality obtained in passive scenarios on ratings after interactive scenarios, the latter provide a scenario-based context that increases not only conversational aspects of the (synthetic) talker, but also a specific context that is relevant for the ultimate preference and acceptance in an application domain (see last section). For evaluation of speech synthesis quality, not only Talker Quality, it is argued to move from passive to simple interactive scenarios in order to define a social context [365]. In that position paper, it is furthermore argued to apply behavioral measures, such as reaction and executions times, and (continuous) physiological measures for relying on multiple sources of synthesis quality, all indicators of what is defined in this book as Talker Quality. Particularly, it is suggested in [365] for allowing participants of such interactive scenarios to show a preference in a comparison of two systems or versions by appropriate scenarios, e.g., by taking advice from either one or the other tutoring system. This challenge is to define scenarios that result in such user choices that relate to the intended evaluation concept. While the selection of a particular agent for a social conversation indicates its assignment to liking and Talker Quality, it still has to be confirmed in laboratory tests by accompanying ratings. For certain task-based scenarios, this might prove to be more difficult. For example, preferring to interact with a specific agent might not be caused by higher Talker Quality, but instead be based solely on attributed competence for the task to be solved, or on attributed trustworthiness of the information provided in a game [222]. Therefore, it has to be confirmed, whether this aspect of competence or trustworthiness is the major quality aspect of Talker Quality or any other intended concept in this particular situation. Apart from the issue of ensuring the status of the behavioral (or even physiological) measure, binary choices in the case of preference provide a robust, but crude response, as frequencies have to be analyzed to describe and model systematics that comprise more conditions than just two. However, this approach can directly be transferred to field tests, as behavior, in terms of durations, errors, and preference choices, can be observed without obstruction, while rating scales, interviews, and especially physiological measures induce at least much organizational effort. Therefore, instead of applying complex implicit laboratory methods or longer questionnaires, simple scales (Sect. 2.1) or even behavioral indication of Talker Quality are more suitable for interactive scenarios in the field. The affective response and behavioral reaction are considered as fundamental part of attitudes, after all. The positive characteristic of noticing behavioral user reactions on agents is the absence of a potential social bias in and the refusing of reporting explicit Talker Quality on a scale.

Acoustic models developed with results from passive scenarios are promising to be applicable as estimators, for example, for perceptual dimensions, or directly for a first impression of Talker Quality. In contrast to this, quantitative models for Talker Quality in interactive scenarios show severe limitations concerning their ability to be used for other conditions than tested. This is apparently caused by interface singularities, but also by the social situation elicited in an interaction. This does not render empirical results useless for other situations and systems, though. While it necessitates empirical evaluation for a new system, the identified dimensions and occurring attributions, with their impact on forming Talker Quality, will not change unexpectedly. Models can thus be used for specifying system characteristics, and for interpreting new user research results, as one basic determiner of the mapping between perceptual dimensions and attributions is the separation between internal and external causes of observed behavior (Sect. 1.2).

But even within a specific situation, such as task-based information systems for a given service domain, prediction models of Talker Quality have an important application domain. For speech dialog managers, reinforcement learning provides an approach to replace hand-crafted dialog behavior by experts [250, 277], which require a lot of effort and are difficult to keep consistent and to maintain for larger services. By defining statistical representations of dialog behavior, which is trained by user simulation, simulated sequences of system actions in conversation are evaluated and used for updating the system strategy [299]. Traditionally, evaluation criteria are simple and based on simple interaction parameters such as task success and number of turns, but more recent approaches successfully incorporated model estimates [339], and particularly user-specific quality estimates [354]. Hence, Talker Quality models can be applied as evaluation criterion not only for predicting data from similar system or interactions, but productively used to develop better systems of the same domain. With more progress on the impact of the first impression on conversational behavior, user simulation could become even more realistic for such purposes, as socially appropriate behavior would be available for conversational agents.

Appendix A
Questionnaire Assessing Descriptors of Voice and Speaking Style

See Table A.1.

Table A.1 The list of labels to describe voice and speaking style

Antonyms (German)	English translation
Klangvoll/klanglos	Sonorous/soft
Tief/hoch	Low/high
Nasal/nicht nasal	Nasal/not nasal
Stumpf/scharf	Smooth/harsh
Gleichmäßig/ungleichmäßig	Even/uneven
Akzentfrei/mit Akzent	Accented/without accent
Dunkel /hell	Dark/bright
Leise/laut	Quiet/loud
Knarrend/nicht knarrend	Creaky/not creaky
Variabel/monoton	Variable/monotonous
Angenehm/unangenehm	Pleasant/unpleasant
Deutlich/undeutlich	Articulate/inarticulate
Rau/glatt	Rough/flat
Klar/heiser	Clear/hoarse
Unauffällig/auffällig	Ordinary/outstanding
Schnell/langsam	Quick/slow
Kalt/warm	Cold/warm
Unnatürlich/natürlich	Unnatural/natural
Stabil/zittrig	Stable/shaky
Unpräzise/präzise	Imprecise/precise
Brüchig/fest	Brittle/firm
Unmelodisch/melodisch	Tuneless/tuneful
Angespannt/entspannt	Tense/relaxed
Holprig/gleitend	Bumpy/gliding

(continued)

B. Weiss, *Talker Quality in Human and Machine Interaction*, T-Labs Series in Telecommunication Services, https://doi.org/10.1007/978-3-030-22769-2

Table A.1 (continued)

Antonyms (German)	English translation
Lang/kurz	Long/short
Locker/gepresst	Lax/pressed
Kraftvoll/kraftlos	Powerful/powerless
Flüssig/stockend	Fluent/halting
Weich/hart	Soft/hard
Professionell/unprofessionell	Professional/unprofessional
Betont/unbetont	Accentuated/unaccented
Sanft/schrill	Gentle/sharp
Getrennt/verbunden	Disconnected/connected
Nicht behaucht/behaucht	Not Aspirated/aspirated

Appendix B
Questionnaire Assessing Attributions of Talkers' Traits and States

See Table B.1.

Table B.1 The list of labels to describe talkers' traits and states

Antonyms (German)	English translation
Sympathisch/unsympathisch	Likable/non-likable
Unsicher/sicher	Insecure/secure
Unattraktiv/attraktiv	Unattractive/attractive
Verständnisvoll/verständnislos	Sympathetic/unsympathetic
Entschieden/unentschieden	Decided/indecisive
Aufdringlich/unaufdringlich	Obtrusive/unobtrusive
Nah/distanziert	Close/distant
Interessiert/gelangweilt	Interested/bored
Emotionslos/emotional	Unemotional/emotional
Genervt/nicht genervt	Irritated/not irritated
Passiv/aktiv	Passive/active
Unangenehm/angenehm	Unpleasant/pleasant
Charaktervoll/charakterlos	Characterful/characterless
Reserviert/gesellig	Reserved/sociable
Nervös/entspannt	Nervous/relaxed
Distanziert/mitfühlend	Distant/affectionate
Unterwürfig/dominant	Conformable/dominant
Affektiert/unaffektiert	Affected/unaffected
Gefühlskalt/herzlich	Cold/hearty
Jung/alt	Young/old
Sachlich/unsachlich	Factual/not factual
Aufgeregt/ruhig	Excited/calm
Kompetent/inkompetent	Competent/incompetent
Schön/hässlich	Beautiful/ugly
Unfreundlich/freundlich	Unfriendly/friendly

(continued)

© Springer Nature Switzerland AG 2020
B. Weiss, *Talker Quality in Human and Machine Interaction*, T-Labs Series
in Telecommunication Services, https://doi.org/10.1007/978-3-030-22769-2

Table B.1 (continued)

Antonyms (German)	English translation
Weiblich/männlich	Feminine/masculine
Provokativ/gehorsam	Offensive/submissive
Engagiert/gleichgültig	Committed/indifferent
Langweilig/interessant	Boring/interesting
Folgsam/zynisch	Compliant/cynical
Unaufgesetzt/aufgesetzt	Genuine/artificial
Dumm/intelligent	Stupid/intelligent
Erwachsen/kindlich	Adult/childish
Frech/bescheiden	Bold/modest

Appendix C
List of Extracted Interaction Parameters From the Annotated Three-Person Conversations for Sect. 3.2.2.2

1. Total time of S turns
2. Total time of whole conversation
3. Averaged turn length of S
4. Percentage of S speaking
5. Total time of overlap (more than one speaker speaking (without back-channel)
6. Number of stretches with overlap (without back-channel)
7. Average duration of S turns (mean)
8. Number of S turns
9. Number of all speakers' turns
10. Percentage of turns to all turns
11. Number of S pauses between speaker's turns (without back-channel)
12. Number of S pauses per turn
13. Percentage of S pauses to all pauses
14. Percentage of S pauses to S time of speaking
15. Number of turn-taking attempts of another while S speaking
16. Number of S turns ending with a pause
17. Percentage of nr number of S turns ending with a pause to nr of S turns
18. Percentage of S turns ending with a pause to all ending with a pause
19. Number of S turns starting with a pause
20. Percentage of nr number of S turns starting with a pause to nr of S turns
21. Percentage of S turns starting with a pause to all ending with a pause
22. Number of S turns ending with overlap
23. Percentage of S turns ending with overlap to nr of S turns
24. Percentage of S turns ending with overlap to all turns ending with overlap
25. Percentage of S turns ending with overlap to S turns ending with pause
26. Number of S turns starts with overlap
27. Percentage of S turns starts with overlap to nr of S turns
28. Percentage of S turns starts with overlap to all turns ending with overlap
29. Percentage of S turns starts with overlap to S turns ending with pause

© Springer Nature Switzerland AG 2020
B. Weiss, *Talker Quality in Human and Machine Interaction*, T-Labs Series
in Telecommunication Services, https://doi.org/10.1007/978-3-030-22769-2

30. Number of S back-channels
31. Percentage of back-channels of S to all back-channels
32. Percentage of back-channels of S to nr of S turns
33. Number of turn-taking attempts of others while S is speaking
34. Percentage of turn-taking attempts of others to S speaking duration
35. Percentage of turn-taking attempts of others to S nr of turns

Appendix D
List of Interaction Parameters Relative to the Evaluating Talker for Sect. 3.2.2.4

1. Ratio of mean duration of S turns to those of Sx
2. Ratio of total time of S turns to those of Sx
3. Ratio of number of S turns to those of Sx
4. Ratio of number of S pauses within the turns to those of Sx
5. Difference of non-successful interruptions (can be zero for some Sx)

© Springer Nature Switzerland AG 2020
B. Weiss, *Talker Quality in Human and Machine Interaction*, T-Labs Series in Telecommunication Services, https://doi.org/10.1007/978-3-030-22769-2

Appendix E
List of Interaction Parameters That Are Dependent on the Evaluating Talker for Sect. 3.2.2.4

1. Number of doubletalk for both (without back-channel)
2. Duration of doubletalk for both (without back-channel)
3. Number of turn changes with pauses for both (without back-channel)
4. Number of turn changes with overlap for both (without back-channel)
5. Number of back-channel

© Springer Nature Switzerland AG 2020

B. Weiss, *Talker Quality in Human and Machine Interaction*, T-Labs Series
in Telecommunication Services, https://doi.org/10.1007/978-3-030-22769-2

References

1. Abele, A.E., Cuddy, A.J.C., Judd, C.M., Yzerbyt, V.Y.: Fundamental dimensions of social judgment. Editorial to the special issue. Eur. J. Soc. Psychol. **38**(7), 1063–1065 (2008)
2. Adcock, A., Van Eck, R.: Reliability and factor structure of the attitude toward tutoring agent scale (ATTAS). J. Interact. Learn. Res. **16**, 195–217 (2005)
3. Ajzen, I., Fishbein, M.: Understanding Attitudes and Predicting Social Behavior. Prentice-Hall, Englewood Cliffs (1980)
4. Albert, W., Gribbons, W., Almadas, J.: Pre-conscious assessment of trust: a case study of financial and health care web sites. In: Proceedings of the Human Factors and Ergonomics Society Annual Meeting, San Antonio, pp. 449–453 (2009)
5. Ambady, N., Bernieri, F.J., Richeson, J.A.: Toward a histology of social behavior: judgmental accuracy from thin slices of the behavioral stream. In: Zanna, M.P. (ed.) Advances in Experimental Social Psychology, vol. 32, pp. 201–272. Academic Press, San Diego (2000)
6. Ambady, N., Rosenthal, R.: Half a minute: predicting teacher evaluations from thin slices of nonverbal behavior and physical attractiveness. J. Pers. Soc. Psychol. **64**, 431–441 (1993)
7. Ambady, N., Skowronski, J.J. (eds.): First Impressions. Guilford Press, New York (2008)
8. Ambady, N., Krabbenhoft, M.A., Hogan, D.: The 30-sec sale: using thin slice judgments to evaluate sales effectiveness. J. Consult. Psychol. **16**, 4–13 (2006)
9. Anderson, A., Bader, M., Bard, E., Boyle, E., Doherty, G.M., Garrod, S., Isard, S., Kowtko, J., McAllister, J., Miller, J., Sotillo, C., Thompson, H.S., Weinert, R.: The HCRC map task corpus. Lang. Speech **34**, 351–366 (1991)
10. Antons, J.N., Arndt, S., Schleicher, R., Möller, S.: Brain activity correlates of quality of experience. In: Möller, S., Raake, A. (eds.) Quality of Experience: Advanced Concepts, Applications and Methods, pp. 109–119. Springer, Heidelberg (2014)
11. Apicella, C., Feinberg, D.: Voice pitch alters mate-choice-relevant perception in hunter-gatherers. Proc. R. Soc. B Biol. Sci. **276**, 1077–1082 (2009)
12. Apple, W., Streeter, L.A., Krauss, R.M.: Effects of pitch and speech rate on personal attributions. J. Pers. Soc. Psychol. **37**(5), 715–727 (1979)
13. Argyle, M.: Bodily Communication. Methuen, New York (1988)
14. Argyle, M., Dean, J.: Eye-contact, distance and affiliation. Sociometry **28**, 289–304 (1965)
15. Aronson, E., Wilson, T., Akert, R.M.: Social Psychology, 7th edn. Prentice Hall (2009)
16. Bachorowski, J.A., Owren, M.: Sounds of emotion: Production and perception of affect-related vocal acoustics. Ann. N.Y. Acad. Sci. **1000**, 244–265 (2003)
17. Back, M.D., Schmukle, S.C., Egloff, B.: A closer look at first sight: social relations lens model analysis of personality and interpersonal attraction at zero acquaintance. Eur. J. Personal. **25**, 225–238 (2011)

© Springer Nature Switzerland AG 2020
B. Weiss, *Talker Quality in Human and Machine Interaction*, T-Labs Series in Telecommunication Services, https://doi.org/10.1007/978-3-030-22769-2

18. Bødker, S.: Third-wave HCI, 10 years later—participation and sharing. Interactions **22**, 24–31 (2015)
19. Bailenson, J.N., Blascovich, J., Beall, A.C., M., J.: Equilibrium theory revisited: Mutual gaze and personal space in virtual environment. Presence **10**, 583–598 (2001)
20. Bailly, G., Amélie, L.: Speech dominoes and phonetic convergence. In: Proceedings of the conference on Interspeech, pp. 1153–1156 (2010)
21. Baker, A., Ayres, J.: The effect of apprehensive behavior on communication apprehension and interpersonal attraction. Commun. Res. Rep. **11**, 45–51 (1994)
22. Baker, R., Hazan, V.: DiapixUK: task materials for the elicitation of multiple spontaneous speech dialogs. Behav. Res. Methods **43**, 761–770 (2011)
23. Bar, M., Neta, M., Linz, H.: Very first impressions. Emotion **6**(2), 269–278 (2006)
24. Baraković, S., Skorin-Kapov, L.: Survey of research on quality of experience modelling for web browsing. Quality User Experience **2:6**, 1–31 (2017)
25. Bartneck, C., Croft, E., Kulic, D., Zoghbi, S.: Measurement instruments for the anthropomorphism, animacy, likeability, perceived intelligence, and perceived safety of robots. Int. J. Soc. Robot. **1**, 71–81 (2009)
26. Baumann, T.: Incremental spoken dialogue processing: Architecture and lower-level components. Ph.D. thesis, University of Bielefeld (2013)
27. Bayerisches Archiv für Sprachsignale: PhonDat 1 (1995)
28. Bell, L., Gustafson, J., Heldner, M.: Prosodic adaption in human-computer interaction. In: Proceedings of ICPHS, pp. 2453–2456 (2003)
29. Bem, D.: Beliefs, attitudes, and human affairs. Brooks/Cole, Belmont (1970)
30. Berg, J., Rumsey, F.: Spatial attribute identification and scaling by repertory grid technique and other methods. In: Proceedings of the AES 16th International Conference on Spatial Sound Reproduction (1999)
31. Berger, C.R., Clatterbuck, G.W.: Attitude similarity and attributional information as determinants of uncertainty reduction and interpersonal attraction. In: Annual Conference of the International Communication Association, Portland (1976)
32. Bergmann, K., Eyssel, F., Kopp, S.: A second chance to make a first impression? How appearance and nonverbal behavior affect perceived warmth and competence of virtual agents over time. In: Proceedings of the Conference on Intelligent Virtual Agents, pp. 126–138. Springer, Berlin (2012)
33. Bernsen, N.O.: From theory to design support tool. In: Ruttkay, Z., Pelachaud, C. (eds.) Multimodality in Language and Speech Systems, pp. 93–148. Kluwer, Dordrecht (2002)
34. Bernsen, N.O., Dybkjær, H., Dybkjær, L.: Cooperativity in human-machine and human-human spoken dialogue. Discourse Process. **21**, 213–236 (1996)
35. Bernsen, N., Dybkjær, L.: Multimodal Usability. Springer, London (2009)
36. Betz, S., Carlmeyer, B., Wagner, P., Wrede, B.: Interactive hesitation synthesis: Modelling and evaluation. Multimodal Technol. Interact **2**, 1–21 (2018)
37. Birkholz, P., Martin, L., Xu, Y., Scherbaum, S., Neuschaefer-Rube, C.: Manipulation of the prosodic features of vocal tract length, nasality and articulatory precision using articulatory synthesis. Comput. Speech Lang. **41**, 116–127 (2017)
38. Borg, I., Groenen, P.: Modern Multidimensional Scaling: Theory and Applications, 2nd edn. Springer, New York (2005)
39. Bortz, J., Schuster, C.: Statistik für Human- und Sozialwissenschaftler, 10th edn. Springer, Berlin (2010)
40. Bose, I.: dóch da sín ja ' nur mûster – Kindlicher Sprechausdruck im sozialen Rollenspiel. No. 9 in Hallesche Schriften zur Sprechwissenschaft und Phonetik. Peter Lang, Frankfurt am Main (2003)
41. Brückl, M.: Altersbedingte Veränderungen der Stimme und Sprechweise von Frauen. Dissertation, Technische Universität Berlin, Berlin (2011)
42. Bradac, J., Mulac, A., House, A.: Lexical diversityand magnitude of convergent versus divergent style shifting perceptual and evaluative consequences. Lang. Commun. **8**, 213–228 (1988)

43. Brandt, D.: On liking social performance with social competence: some relations between communicative and attributions of interpersonal attractiveness and effectiveness. Hum. Commun. Res. **5**, 223–226 (1979)

44. Branigan, H.P., Pickering, M.J., Pearson, J., Mclean, J.F.: Linguistic alignment between people and computers. J. Pragmat. **42**, 2355–2368 (2010)

45. Brennan, S.E., Clark, H.H.: Lexical choice and conceptual pacts in conversation. J. Exp. Psychol. Learn. Mem. Cogn. **11**, 1482–1493 (1996)

46. Brockmann, C., Isard, A., Oberlander, J., White, M.: Modelling alignment for affective dialogue. In: Proceedings of the Workshop on Adapting the Interaction Style to Affective Factors at the 10th International Conference on User Modeling, pp. 1–5 (2005)

47. Brown, B., Giles, H., Thakerar, J.: Speaker evaluation as a function of speech rate, accent, and context. Lang. Commun. **5**(3), 207–220 (1985)

48. Brown, B.L., Strong, W.J., Rencher, A.C.: Perceptions of personality from speech: effects of manipulations of acoustical parameter. J. Acoust. Soc. Am. **54**(1), 29–35 (1973)

49. Brown, B.L., Strong, W.J., Rencher, A.C.: Fifty-four voices from two: the effects of simultaneous manipulations of rate, mean fundamental frequency, and variance of fundamental frequency on ratings of personality from speech. J. Acoust. Soc. Am. **55**(2), 313–318 (1974)

50. Brown, B.L., Strong, W.J., Rencher, A.C.: Acoustic determinants of perceptions of personality from speech. Linguistics **13**(166), 11–32 (1975)

51. Bruckert, L., Lienard, J., Lacroix, A., Kreutzer, M., Leboucher, G.: Women use voice parameter to assess men's characteristics. Proc. Biol. Sci. **237**(1582), 83–89 (2006)

52. Burgoon, J.K.: Attributes of the newscaster's voice as predictors of his credibility. Journal. Q. **55**, 276–281 (1978)

53. Burgoon, J., Buller, D., Hale, J., Turck, M.: Relational messages associated with nonverbal behaviors. Hum. Commun. Res. **10**, 351–378 (1984)

54. Burkhardt, F., Schuller, B., Weiss, B., Weninger, F.: Would you buy a car from me? On the likability of telephone voices. In 12th Interspeech, Florence, pp. 1557–1560 (2011)

55. Burkhardt, F., Paeschke, A., Rolfes, M., Sendlmeier, W., Weiss, B.: A database of German emotional speech. In: Proceedings of the 6th Interspeech Conference, Lisbon, pp. 1517–1520 (2005)

56. Burkhardt, F., Weiss, B., Eyben, F., Deng, J., Schuller, B.: Detecting vocal irony. In: Proceedings of the Conference on German Society for Computational Linguistics and Language Technology, pp. 16–191 (2017)

57. Burmester, M., Mast, M., Jäger, K., Homans, H.: Valence method for formative evaluation of user experience. In: Proceedings of the 8th ACM Conference on Designing Interactive Systems, pp. 364–367. ACM, New York (2010)

58. Burton, M.L., Nerlove, S.B.: Balanced designs for triads tests: two examples from English. Soc. Sci. Res. **5**, 247–267 (1976)

59. Buschmeier, H., Bergmann, K., Kopp, S.: An alignment-capable microplanner for natural language generation. In: Proceedings of the 12th European Workshop on Natural Language Generation, p. 82–89. ACM, New York (2007)

60. Byrne, D.: The Attraction Paradigm. Academic Press, San Diego (1971)

61. Cacioppo, J.T., Crites, S.L.J., Gardner, W.L., Berntson, G.G.: Bioelectrical echoes from evaluative categorizations: I. A late positive brain potential that varies as a function of trait negativity and extremity. J. Pers. Soc. Psychol. **67**, 115–125 (1994)

62. Cafaro, A., Vilhjálmsson, H., Bickmore, T.: First impressions in human–agent virtual encounters. ACM Trans. Comput. Hum. Interact. **23**, 24:1–40 (2016)

63. Calder, A., Rhodes, G., Johnson, M., Haxby, J. (eds.): Oxford Handbook of Face Perception. Oxford University Press, Oxford (2011)

64. Callejas, Z., Ravenet, B., Ochs, M., Pelachaud, C.: A computational model of social attitudes for a virtual recruiter. In: Proceedings of the International Conference on Autonomous Agents and Multiagent Systems (AAMAS), pp. 93–100 (2014)

65. Castro-González, A., Admoni, H., Scassellati, B.: Effects of form and motion on judgments of social robots' animacy. Int. J. Hum. Comput. Stud. **90**, 27–38 (2016)

66. Chartrand, T.L., Bargh, J.A.: The chameleon effect: The perception-behavior link and social interaction. J. Pers. Soc. Psychol. **76**(6), 893–910 (1999)
67. Chattopadhyay, A., Dahl, D.W., Ritchie, R.J., Shahin, K.N.: Hearing voices: the impact of announcer speech characteristics on consumer response to broadcast advertising. J. Consum. Psychol. **13**(3), 198–204 (2003)
68. Coan, J.A., Allen, J.J.: Frontal EEG asymmetry as a moderator and mediator of emotion. Biol. Psychol. **67**, 7–49 (2004)
69. Collins, S.: Men's voices and women's choices. Anim. Behav. **60**(6), 773–780 (1993)
70. Collins, S., Missing, C.: Vocal and visual attractiveness are related in women. Anim. Behav. **65**(5), 997–1004 (2003)
71. Costello, A.B., Osborne, J.W.: Exploratory factor analysis: four recommendations for getting the most from your analysis. Pract. Assess. Res. Eval. **10**, 1–9 (2005)
72. Cowan, B., Branigan, H., Obregón, M., Bugis, E., Beale, R.: Voice anthropomorphism, interlocutor modelling and alignment effects on syntactic choices in human-computer dialogue. Int. J. Hum. Comput. Stud. **83**, 27–42 (2015)
73. Cuayáhuitl, H., Renals, S., Lemon, O., Shimodaira, H.: Evaluation of a hierarchical reinforcement learning spoken dialogue system. Comput. Speech Lang. **24**, 395–429 (2010)
74. Cuddy, A.J., Fiske, S.T., Glick, P.: Warmth and competence as universal dimensions of social perception: the stereotype content model and the bias map. Adv. Exp. Soc. Psychol. **40**, 62–149 (2008)
75. Cunningham, W.A., Espinet, S.D., DeYoung, C.G., Zelazo, P.D.: Attitudes to the right- and left: frontal erp asymmetries associated with stimulus valence and processing goals. NeuroImage **28**, 827–834 (2005)
76. Curhan, J.R., Pentland, A.: Thin slices of negotiation: Predicting outcomes from conversational dynamics within the first 5 minutes. J. Appl. Psychol. **92**(3), 802–811 (2007)
77. Dabbs, J.M.: Similarity of gestures and interpersonal influence. In: Proceedings of the Annual Convention of the American Psychological Association, vol. 4, pp. 337–338 (1969)
78. Davies, B.: Grice's cooperative principle: getting the meaning across. Leeds working papers in linguistics, University of Leeds (2008)
79. Davis, F.: Perceived usefulness, perceived ease of use, and user acceptance of information technology. MIS Q. **13**, 319–340 (1989)
80. de Jong, N.H., Wempe, T.: Praat script to detect syllable nuclei and measure speech rate automatically. Behav. Res. Methods **41**, 385–390 (2009)
81. De Looze, C., Scherer, S., Vaughan, B., Campbell, N.: Investigating automatic measurements of prosodic accommodation and its dynamics in social interaction. Speech Commun. **58**, 11–34 (2014)
82. DeGroot, T., Motowidlo, S.J.: Why visual and vocal interview cues can affect interviewers' judgments and predict job performance. J. Appl. Psychol. **84**(6), 986–993 (1999)
83. Dehn, D., van Mulken, S.: The impact of animated interface agents: a review of empirical research. Int. J. Hum. Comput. Stud. **52**, 1–22 (2000)
84. Demaree, H.A., Everhart, D.E., Youngstrom, E.A., Harrison, D.W.: Brain lateralization of emotional processing: historical roots and a future incorporating "dominance". Behav. Cogn. Neurosci. Rev. **4**, 3–20 (2005)
85. Demeure, V., Niewiadomski, R., Pelachaud, C.: How believability of virtual agent is linked to warmth, competence, personification and embodiment? Presence **20**(5), 431–448 (2011)
86. DePaulo, B.M., Kenny, D.A., Hoover, C.W., Webb, W., Oliver, P.V.: Accuracy of person perception: Do people know what kinds of impressions they convey? J. Pers. Soc. Psychol. **52**(2), 303–315 (1987)
87. Diefenbach, S., Hassenzahl, M.: Handbuch zur fun-ni toolbox. user experience evaluation auf drei ebenen. Tech. rep., Folkwang Universität (2010). http://fun-ni.org/wp-content/uploads/Diefenbach+Hassenzahl_2010_HandbuchFun-niToolbox.pdf
88. Dohen, M.: Speech through the ear, the eye, the mouth and the hand. In: Esposito, A., Hussain, A., Marinaro, M. (eds.) Multimodal Signals: Cognitive and Algorithmic Issues. Springer, Berlin (2009)

89. Duran, D., Lewandowski, N., Bruni, J., Schweitzer, A.: Akustische Korrelate wahrgenommener Persönlichkeitsmerkmale und Stimmattraktivität. In: Proceedings of the Elektronische Sprachsignalverarbeitung, pp. 91–98. TUD Press, Dresden (2017)

90. Dybkjær, L., Bernsen, N.O., Dybkjær, H.: Grice incorporated. Cooperativity in spoken dialogue. In: Proceedings of COLING, pp. 328–333 (1996)

91. Eagly, A.H., Chaiken, S.: The Psychology of Attitudes. Harcourt Brace Jovanovich, Fort Worth (1993)

92. Eckert, H., Laver, J.: Menschen und ihre Stimmen: Aspekte der vokalen Kommunikation. Beltz, Weinheim (1994)

93. Ehrenbrink, P., Möller, S.: Development of a reactance scale for human–computer interaction. Quality User Experience **3:2**, 1–13 (2018)

94. Enfield, N.J.: How we talk. The Inner Workings of Conversation. Basic Books, New York (2017)

95. Engelbrecht, K.P., Kühnel, C., Möller, S.: Weighting the coefficients in PARADISE models to increase their generalizability. In: André, E. Dybkjær, L., Minker, W., Neumann, H., Pieraccini, R., Weber, M. (eds.) 4th IEEE Workshop on Perception and Interactive Technologies for Speech-Based Systems (PIT), Kloster Irsee, LNAI 5078, pp. 289–292. Springer, Berlin (2008)

96. Estival, D., Cassidy, S., Cox, F., Burnham, D.: AusTalk: an audio-visual corpus of Australian English. In: Proceedings of the Language Resources Evaluation Conference (LREC), pp. 3105–3109 (2014)

97. Evanini, K., Hunter, P., Liscombe, J., Sündermann, D., Dayanidhi, K., Pieraccini, R.: Caller experience: a method for evaluating dialog systems and its automatic prediction. In: Proceedings of the Spoken Language Technology Workshop, SLT, pp. 129–132 (2008)

98. Fagel, W., Herpt, L.V.: Analysis of the perceptual qualities of Dutch speakers' voice and pronunciation. Speech Commun. **1**, 315–326 (1983)

99. Fandrianto, A., Eskenazi, M.: Prosodic entrainment in an information-driven dialog system. In: Proceedings of the Interspeech, pp. 1–4 (2012)

100. Förster, J., Strack, F.: Motor actions in retrieval of valenced information: II. Boundary conditions for motor congruence effects. Percept. Mot. Skills **86**, 1423–1426 (1998)

101. Fazio, R.H., Sanbonmatsu, D.M., Powell, M.C., Kardes, F.R.: On the automatic activation of attitudes. J. Pers. Soc. Psychol. **50**, 229–238 (1986)

102. Feinberg, D.: Are human faces and voices ornaments signaling common underlying cues to mate value? Evol. Anthropol. Issues News Rev. **17**(2), 112–118 (2008)

103. Feinberg, D., DeBruine, L., Jones, B., Perrett, D.: The role of femininity and averageness of voice pitch in aesthetic judgments of women's voices. Perception **37**, 615–623 (2008)

104. Feldstein, S., Dohm, F., Crown, C.: Gender and speech rate in the perception of competence and social attractiveness. J. Soc. Psychol. **141**, 785–806 (2001)

105. Ferdenzi, C., Patel, S., Mehu-Blantar, I., Khidasheli, M., Sander, D., Delplanque, S.: Voice attractiveness: influence of stimulus duration and type. Behav. Res. Methods **45**, 405–413 (2013)

106. Ferguson, M.J., Fukukura, J.: Likes and dislikes: A social cognitive perspective on attitudes. In: Fiske, S.T., Macrae, C.N. (eds.) The SAGE Handbook of Social Cognition, pp. 165–186. SAGE Publications, London (2012)

107. Fernández Gallardo, L.: The Nautilus Speaker Characterization Corpus (2018). http://hdl.handle.net/11022/1009-0000-0007-C05F-6. ISLRN: 157-037-166-491-1

108. Fernández Gallardo, L., Weiss, B.: Speech likability and personality-based social relations: a round-robin analysis over communication channels. In: Interspeech, pp. 903–907 (2016)

109. Fernández Gallardo, L., Weiss, B.: Perceived interpersonal speaker attributes and their acoustic features. In: 13th Phonetik & Phonology im deutschsprachigem Raum, pp. 61–64 (2017)

110. Fernández Gallardo, L., Weiss, B.: Towards speaker characterization: identifying and predicting dimensions of person attribution. In: Interspeech, pp. 904–908 (2017)

111. Fernández Gallardo, L., Weiss, B.: The Nautilus speaker characterization corpus: speech recordings and labels of speaker characteristics and voice descriptions. In: Proceedings of the LREC, pp. 2837–2842 (2018)

112. Fiske, S., Pavelchak, M.: Category-based versus piecemeal-based affective responses: developments in schema-triggered affect. In: Sorrentino, R., Higgins, E. (eds.) The Handbook of Motivation and Cognition: Foundation of Social Behaviour, pp. 167–203. Guilford Press, New York (1986)

113. Fiske, S.T., Cuddy, A.J., Glick, P.: Universal dimensions of social cognition: warmth and competence. Trends Cogn. Sci. **11**(2), 77–83 (2006)

114. Fitch, W., Giedd, J.: Morphology and development of the human vocal tract: a study using magnetic resonance imaging. J. Acoust. Soc. Am. **106**, 1511–1522 (1999)

115. Floyd, K., Erbert, L.: Relational message interpretations of nonverbal matching behavior: an application of the social meaning model. J. Soc. Psychol. **143**, 581–597 (2003)

116. Foster, M.: Enhancing human-computer interaction with embodied conversational agents. In: Proceedings of the International Conference on Universal Access in Human-Computer Interaction: Ambient Interaction, pp. 828–837 (2007)

117. Foster, M., Giuliani, M., Knoll, A.: Comparing objective and subjective measures of usability in a human-robot dialogue system. In: Proceedings of the International Conference on Universal Access in Human-Computer Interaction: Ambient Interaction, pp. 879–887 (2009)

118. Fransella, F., Bell, R., Bannister, D.: A Manual for Repertory Grid Technique, 2nd edn. Wiley, Chichester (2004)

119. Gödde, F., Möller, S., Engelbrecht, K.P., Kühnel, C., Schleicher, R., Naumann, A., Wolters, M.: Study of a speech-based smart home system with older users. In: Proceedings of the International Workshop on Intelligent User Interfaces for Ambient Assisted Living, pp. 17–22 (2008)

120. Gövercin, M., Meyer, S., Schellenbach, M., Steinhagen-Thiessen, E., Weiss, B., Haesner, M.: Smartsenior@home: Acceptance of an integrated ambient assisted living system. results of a clinical field trial in 35 households. Inform. Health Soc. Care **41**, 1–18 (2016)

121. Georgakopoulos, I.: Comparing explicit and implicit interpersonal attraction based on auditory impressions of unacquainted speakers. Bachelor thesis, Technische Universität Berlin (2016)

122. Gibbon, D., Mertins, I., Moore, R. (eds.): Handbook of Multimodal and Spoken Dialogue Systems: Resources, Terminology and Product Evaluation. Kluwer, Norwell (2000)

123. Gilbert, D., Malone, P.: The correspondence bias. Psychol. Bull. **17**, 21–38 (1995)

124. Giles, H.: Accommodation theory: some new directions. York Papers Linguist. **9**, 105–136 (1980)

125. Goldbrand, S.: Imposed latencies, interruptions and dyadic interaction: physiological response and interpersonal attraction. J. Res. Pers. **15**, 221–232 (1981)

126. Goldstein, E., Brockmole, J.: Sensation and Perception, 10th edn. Cengage Learning, Boston (2016)

127. Gottman, J., Notarius, C.: Decade review: observing marital interaction. Interactions **62**, 927–947 (2000)

128. Goudbeek, M., Scherer, K.: Beyond arousal: Valence and potency/control cues in the vocal expression of emotion. J. Acoust. Soc. Am. **128**(3), 1322–1336 (2010)

129. Gravano, A.: Turn-taking and affirmative cue words in task-oriented dialogue. Ph.D. thesis, Columbia University (2009)

130. Gravano, S., Beňuš, Š., Levitan, R., Hirschberg, J.: Backward mimicry and forward influence in prosodic contour choice in standard American English. In: Proceedings of the Interspeech, pp. 1839–1843 (2015)

131. Gravano, A., Levitan, R., Willson, L., Beňuš, Š., Hirschberg, J., Nenkova, A.: Acoustic and prosodic correlates of social behavior. In: Proceedings of the Interspeech, pp. 97–100 (2011)

132. Greenwald, A.G., McGhee, D.E., Schwartz, J.L.: Measuring individual differences in implicit cognition: The implicit association test. J. Pers. Soc. Psychol. **74**(6), 1464–1480 (1998)

133. Greenwald, A.G., Nosek, B.A., Banaji, M.R.: Understanding and using the implicit association test: I. An improved scoring algorithm. J. Pers. Soc. Psychol. **85**, 197–216 (2003)
134. Grice, H.P.: Logic and conversation. In: Cole, P., Morgan, J.L. (eds.): Speech Acts, Syntax and Semantics, vol. 3, pp. 41–58. Academic Press, New York (1975)
135. Guerin, B.: Social Facilitation. Cambridge University Press, Cambridge (1993)
136. Gunes, H.: Automatic, dimensional and continuous emotion recognition. Int. J. Synthetic Emotions **1**(1), 68–88 (2010)
137. Hajdinjak, M., Mihelic, F.: The PARADISE evaluation framework: issues and findings. Comput. Linguist. **32**, 263–272 (2006)
138. Hammerberg, B., Fritzell, B., Gauffin, J., Sundberg, J., Wedin, L.: Perceptual and acoustical correlates of abnormal voice qualities. Acta Otolaryngol. **90**, 441–451 (1980)
139. Haris, M., Garris, C.: You never get a second chance to make a first impression. In: Ambady, N., Skowronski, J.J. (eds.): First Impressions, pp. 147–167. Guilford Press, New York (2008)
140. Harmon-Jones, E., Gable, P.A., Peterson, C.K.: The role of asymmetric frontal cortical activity in emotion-related phenomena: A review and update. Biol. Psychol. **84**, 451–462 (2010)
141. Harrison, S., Tatar, D., Sengers, P.: The three paradigms of HCI. In: Proceedings of CHI, pp. 1–18. ACM, New York (2007)
142. Hart, R.J., Brown, B.L.: Personality information contained in the verbal qualities and in content aspects of speech. Speech Monograp. **41**, 271–380 (1974)
143. Hassenzahl, M., Diefenbach, S., Göritz, A.: Needs, affect, and interactive products—facets of user experience. Interacting Comput. **22**, 353–362 (2010)
144. Hassenzahl, M., Monk, A.: The inference of perceived usability from beauty. Hum. Comput. Interact. **25**(3), 235–260 (2010)
145. Hassenzahl, M., Wessler, R.: Capturing design space from a user perspective: the repertory grid technique revisited. Int. J. Hum. Comput. Interact. **12**(3,4), 441–459 (2000)
146. Hecht, M.A., LaFrance, M.: How (fast) can i help you? Tone of voice and telephone operator efficiency in interactions. J. Appl. Soc. Psychol. **25**(23), 2086–2098 (1995)
147. Hedrick, J., Ksiazkiewicz, A.: Getting answers without asking questions: the practicality of an auditory IAT (2011). Draft
148. Heldner, J., Edlund, M., Hirschberg, J.: Pitch similarity in the vicinity of backchannels. In: Proceedings of the Interspeech, pp. 1–4 (2010)
149. Herkner, W.: Lehrbuch Sozialpsychologie. Huber, Bern (1991)
150. Hermann, F., Niedermann, I., Peissner, M., Henke, K., Naumann, A.: Users interact differently: towards a usability-oriented taxonomy. In: Jacko, J. (ed.) Interaction Design and Usability, HCII 2007, No. 4550 in LNAI, pp. 812–817. Springer, Heidelberg (2007)
151. Hinterleitner, F.: Quality of Synthetic Speech: Perceptual Dimensions, Influencing Factors, and Instrumental Assessment. T-Labs Series in Telecommunication Services. Springer, Berlin (2017)
152. Hoßfeld, T., Heegaard, P., Varela, M., Möller, S.: QoE beyond the MOS: an in-depth look at QoE via better metrics and their relation to MOS. Quality User Experience **1**(2), 1–23 (2016)
153. Hoeldtke, K., Raake, A.: Conversation analysis of multi-party conferencing and its relation to perceived quality. In: Proceedings of the International Conference on Communications (ICC), IEEE, pp. 1–5. Kyoto, Japan (2011)
154. Huston, T.L., Levinger, G.: Interpersonal attraction and relationships. Annu. Rev. Psychol. **29**, 115–156 (1978)
155. ISO 24617-2:2012: Language resource management—semantic annotation framework (SemAF), Part 2: Dialogue acts (2012)
156. ISO 9421-11: Ergonomic Requirements for Office Work with Visual Display Terminals (VDTs). Part 11: Guidance on Usability. ISO, Geneva (1998)
157. ISO 9421-210: Ergonomics of human system interaction, Part 210: Human-centred design for interactive systems. ISO, Geneva (2010)
158. ISO/IEC 14496-2: Information technology—coding of audio-visual objects—Part 2: visual. JTC 1/SC 29 coding of audio, picture, multimedia and hypermedia information (2004)

159. ITU-R BS.1116-3: Methods for the Subjective Assessment of Small Impairments in Audio Systems. International Telecommunication Union, Geneva (2015)
160. ITU-R BS.1534-3: Method for the Subjective Assessment of Intermediate Quality Level of Audio Systems. International Telecommunication Union, Geneva (2015)
161. ITU-T Rec. E.800: Definitions of Terms Related to Quality of Service. International Telecommunication Union, Geneva (2008)
162. ITU-T Rec. P.800: Methods for Subjective Determination of Transmission Quality. International Telecommunication Union, Geneva (1996)
163. ITU-T Rec. P.805: Subjective Evaluation of Conversational Quality. International Telecommunication Union, Geneva (2007)
164. ITU-T Rec. P.851: Subjective Quality Evaluation of Telephone Services Based on Spoken Dialogue Systems. International Telecommunication Union, Geneva (2003)
165. ITU-T Suppl. 24 to P-Series: Parameters Describing the Interaction with Spoken Dialogue Systems. International Telecommunication Union, Geneva (2005)
166. ITU-T Suppl. 25 to P-Series: Parameters Describing the Interaction with Multimodal Dialogue Systems. International Telecommunication Union, Geneva (2011)
167. ITU-T Suppl. 26 to P-Series: Scenarios for the Subjective Evaluation of Three-Party Audio Telemeetings Quality. International Telecommunication Union, Geneva (2012)
168. Jokinen, K.: Challenges for adaptive conversational agents. In: Proceedings of the Baltic Conferences on Human Language Technologies, pp. 51–60 (2005)
169. Jokinen, K., Hurtig, T.: User expectations and real experience on a multimodal interactive. In: Proceedings of the Interspeech, pp. 1049–1052 (2006)
170. Jokinen, K., McTear, M.: Spoken Dialogue Systems. Synthesis Lectures on Human-Centered Informatics. Morgan & Claypool, Wadsworth (2010)
171. Jones, E.E.: Interpersonal Perception. Freeman and Company, New York (1990)
172. Kühnel, C.: Quantifying Quality Aspects of Multimodal Interactive Systems. T-Labs Series in Telecommunication Services. Springer, Berlin (2011)
173. Kühnel, C., Weiss, B., Möller, S.: Talking heads for interacting with spoken dialog smart-home systems. In: 10th Interspeech, Brighton, pp. 304–307 (2009)
174. Kühnel, C., Weiss, B., Möller, S.: Evaluating multimodal systems—a comparison of established questionnaires and interaction parameters. In: ACM NordiCHI, Reykjavik, pp. 286–293 (2010)
175. Kühnel, C., Weiss, B., Möller, S.: Parameters describing multimodal interaction—definitions and three usage scenarios. In: 11th Interspeech, Makuhari, pp. 2014–2017 (2010)
176. Kühnel, C., Weiss, B., Schulz, M., Möller, S.: Quality aspects of multimodal dialog systems: identity, stimulation and success. In: 12th Interspeech, Florence, pp. 1349–1352 (2011)
177. Kahneman, D.: Experienced utility and objective happiness: a moment-based approach. In: Choices, Values and Frames, pp. 673–692. Cambridge University Press, New York (2000)
178. Kaplan, K., Firestone, I., Klein, K., Sodikoff, C.: Distancing in dyads: a comparison of four models. Soc. Psychol. Q. **46**, 108–115 (1983)
179. Karacora, B., Dehghani, M., Krämer-Mertens, N., Gratch, J.: The influence of virtual agents' gender and rapport on enhancing math performance. In: Proceedings of the COGSCI, pp. 563–568 (2012)
180. Karnop, C., Weiss, B.: Zum Effekt von Tempo, Tonhöhe und Sprecherformant auf Sympathiebewertungen: Ein Resyntheseexperiment. In: 27th Konferenz Elektronische Sprachsignalverarbeitung (ESSV), Leipzig. Studientexte zur Sprachkommunikation, vol. 81, pp. 206–213. TUD Press, Dresden (2016)
181. Köster, F., Schiffner, F., Möller, S., Malfait, L.: Towards degradation decomposition for voice communication system assessment. Quality User Experience **2:4**, 1–22 (2017)
182. Keizer, S., Kastoris, P., Foster, M.E., Deshmukh, A., Lemon, O.: Evaluating a social multi-user interaction model using a Nao robot. In: Proceedings of the IEEE International Symposium on Robot and Human Interactive Communication, pp. 318–322. IEEE, Piscataway (2014)
183. Kelly, G.: The Psychology of Personal Constructs. Norton, New York (1955)

184. Kennedy, J., Baxter, P., Belpaeme, T.: Comparing robot embodiments in a guided discovery learning interaction with children. Int. J. Soc. Robot. **7**, 293–308 (2014)
185. Kenny, D.: Models of non-independence in dyadic research. J. Soc. Pers. Relat. **13**, 279–294 (1996)
186. Kenny, D.A.: Interpersonal Perception: A social relations analysis. Guilford Press, New York (1994)
187. Kenny, D.A., Kashy, D.A., Cook, W.: Dyadic Data Analysis. Guilford Press, New York (2006)
188. Ketrow, S.: Attributes of a telemarketer's voice and persuasiveness: a review and synthesis of the literature. J. Direct Mark. **4**, 7–21 (1990)
189. Ketzmerick, B.: Zur auditiven und apparativen Charakterisierung von Stimmen. Studientexte zur Sprachkommunikation. TUD Press, Dresden (2007)
190. King, S., Wihlborg, L., Guo, W.: The Blizzard Challenge 2017. In: Proceedings of the Blizzard Challenge Workshop, pp. 1–17 (2017)
191. Kitawaki, N., Itoh, K.: Pure delay effects on speech quality in telecommunications. IEEE J. Sel. Areas Commun. **9**, 586–593 (1991)
192. Knapp, M.L.: Social Intercourse: From Greeting to Goodbye. S. Allyn and Bacon, Needham Heights, MA (1978)
193. Knapp, M., Hall, J.: Nonverbal Communication in Human Interaction. Thomas Learning, Wadsworth (2010)
194. Knapp, M.L., Vangelisti, A.L., Caughlin, J.P.: Interpersonal Communication and Human Relationships, 7th edn. Pearson, London (2014)
195. Kohn, L., Dipboye, R.: The effect on interview structure on recruiting outcomes. J. Appl. Soc. Psychol. **28**, 821–843 (1998)
196. Kolkmeier, J., Vroon, J., Heylen, D.: Interacting with virtual agents in shared space: single and joint effects of gaze and proxemics. In: Proceedings of the Intelligent Virtual Agents, pp. 1–14 (2016)
197. Král, P., Cerisara, C.: Automatic dialogue act recognition with syntactic features. Lang. Resour. Eval. **48**, 419–441 (2016)
198. Krämer, C.: Soziale Wirkung virtueller Helfer. Kohlhammer, Stuttgart (2008)
199. Krämer, N., Hoffmann, L., Kopp, S.: Know your users! empirical results for tailoring an agent's nonverbal behavior to different user groups. In: Proceedings of the Intelligent Virtual Agents, pp. 468–474 (2010)
200. Krämer, N., Kopp, S., Becker-Asano, C., Sommer, N.: Smile and the world will smile with you-the effects of a virtual agent's smile on users' evaluation and behavior. Int. J. Hum. Comput. Stud. **71**, 335–349 (2013)
201. Krämer, N., Simons, N., Kopp, S.: The effects of an embodied agent's nonverbal behaviour on user's evaluation and behavioral mimicry. In: Proceedings of the Intelligent Virtual Agents, pp. 238–251 (2007)
202. Krämer, N.C., Rosenthal-von der Pütten, A.M., Edinger, C.: The effects of a robot's nonverbal behavior on users' mimicry and evaluation. In: Proceedings of the Intelligent Virtual Agents, pp. 442–446 (2016)
203. Krauss, R., Freyberg, R., Morsella, E.: Inferring speakers' physical attributes from their voices. J. Exp. Soc. Psychol. **38**(6), 618–625 (2002)
204. Krause, S., Back, M.D., Egloff, B., Schmukle, S.C.: Implicit interpersonal attraction in small groups automatically activated evaluations predict actual behavior toward social partners. Soc. Psychol. Personal. Sci. **20**, 671–679 (2014)
205. Kreiman, J., Gerratt, B.: The perceptual structure of pathologic voice quality. J. Acoust. Soc. Am. **100**, 1787–1795 (1996)
206. Kreiman, J., Papcun, G.: Comparing discrimination and recognition of unfamiliar voices. Speech Commun. **10**, 265–275 (1991)
207. Krosnick, J.A., Judd, C.M., Wittenbrink, B.: The measurement of attitudes. In: Albarracín, D., Johnson, B.T., Zanna, M.P. (eds.) The Handbook of Attitudes, pp. 21–76. Erlbaum, Mahwah, New York (2005)

208. Kujala, S., Roto, V., Väänänen-Vainio-Mattila, K., Karapanos, E., Sinnelä, A.: UX curve: a method for evaluating long-term user experience. Interact. Comput. **23**(5), 473–483 (2011)

209. Kulviwat, S., II, G.B., Kumar, A., Nasco, S., Clark, T.: Toward a unified theory of consumer acceptance technology. Psychol. Mark. **24**, 1059–1084 (2007)

210. Lai, C., Carletta, J., Renals, S.: Modelling participant affect in meetings with turn-taking features. In: Proceedings of the Workshop of Affective Social Speech Signals (2013)

211. Lakin, J., Jefferis, V., Cheng, C., Chartrand, T.: The chameleon effect as social glue: Evidence for the evolutionary significance of nonconscious mimicry. J. Nonverbal Behav. **27**(3), 145–162 (2003)

212. LaPrelle, J., Hoyle, R., Insko, C., Bernthal, P.: Interpersonal attraction and descriptions of the traits of others: Ideal similarity, self similarity, and liking. J. Res. Pers. **24**, 216–240 (1990)

213. Laver, J.: The Phonetic Description of Voice Quality. University Press, Cambridge (1980)

214. Lavie, T., Tractinsky, N.: Assessing dimensions of perceived visual aesthetics of web sites. Int. J. Hum. Comput. Stud. **60**, 269–298 (2004)

215. Law, E., Roto, V., Hassenzahl, M., Vermeeren, A., Kort, J.: Understanding, scoping and defining User Experience: a survey approach. In: Proceedings of the 27th International Conference on Human Factors in Computing Systems, CHI, pp. 719–728 (2009)

216. Le Callet, P., Möller, S., Perkis, A.: Qualinet white paper on definitions of Quality of Experience, version 1.1, June 3, 2012 (2012). http://www.qualinet.eu/images/stories/whitepaper_v1.1_dagstuhl_output_corrected.pdf. European Network on Quality of Experience in Multimedia Systems and Services (COST Action IC 1003)

217. Lee, C.C., Katsamanis, A., Black, M., Baucom, B., Christensen, A., Georgiou, P., Narayanan, S.S.: Computing vocal entrainment: A signal-derived PCA-based quantification scheme with application to affect analysis in married couple interactions. Comput. Speech Lang. **28**, 518–539 (2013)

218. Lee, D., Lee, J., Kim, E.K., Lee, J.: Dialog act modeling for virtual personal assistant applications using a small volume of labeled data and domain knowledge. In: Proceedings of the Interspeech, p. 1231–1235 (2015)

219. Lee, K.M., Jung, Y., Kim, J., Kim, S.R.: Are physically embodied social agents better than disembodied social agents? the effects of physical embodiment, tactile interaction, and people's loneliness in human-robot interaction. Int. J. Hum. Comput. Stud. **64**, 962–973 (2006)

220. Leino, T., Laukkanen, A.M., Radolf, V.: Formation of the actor's/speakers' formant: a study applying spectrum analysis and computer modeling. J. Voice **25**(2), 150–158 (2011)

221. Levinger, G., Snoek, J.D.: Attraction in Relationship: A New Look at Interpersonal Attraction. General Learning Press, Morristown (1972)

222. Levitan, R.: Acoustic-prosodic entrainment in human-human and human-computer dialogue. Ph.D. thesis, University of Columbia (2014)

223. Levitan, R., Hirschberg, J.: Measuring acoustic-prosodic entrainment with respect to multiple levels and dimensions. In: Proceedings of the Interspeech, pp. 3081–3084. (2011)

224. Levitan, R., Beňuš, S., Gálvez, R., Gravano, A., Savoretti, F., Trnka, M., Weise, A., Hirschberg, J.: Implementing acoustic-prosodic entrainment in a conversational avatar. In: Proceedings of the Interspeech, pp. 1166–1170 (2016)

225. Lewandowski, N., Schweitzer, A.: Prosodic and segmental convergence in spontaneous German conversations. J. Acoust. Soc. Am. **128**, 1458 (2010)

226. Li, J.: The benefit of being physically present: A survey of experimental works comparing copresent robots, telepresent robots and virtual agents. Int. J. Hum. Comput. Stud. **77**, 23–37 (2015)

227. Lindgaard, G., Dudek, C., Sen, D., Sumegi, L., Noonan, P.: An exploration of relations between visual appeal, trustworthiness and perceived usability of homepages. ACM Trans. Comput. Hum. Interact. **18**(1), 1–30 (2011)

228. Lindgaard, G., Fernandes, G., Dudek, C., Brown, J.: Attention web designers: you have 50 milliseconds to make a good first impression! Behav. Inform. Technol. **25**(2), 115–126 (2006)

229. Linville, S.: Vocal Aging. Singular Thomson Learning, San Diego (2001)

230. Lopes, J., Eskenazi, M., Trancoso, I.: Automated two-way entrainment to improve spoken dialog system performance. In: IEEE Conference on Acoustics, Speech and Signal Processing (ICASSP), p. 8372–8376. IEEE, Piscataway (2013)

231. López-Cózar Delgado, R., Araki, M.: Spoken, multilingual and multimodal dialogue systems: development and assessment. Wiley, Chichester (2005)

232. Lubold, N., Pon-Barry, H., Walker, E.: Naturalness and rapport in a pitch adaptive learning companion. In: Proceedings of the IEEE Automatic Speech Recognition and Understanding Workshop, pp. 1–8. IEEE, Piscataway (2015)

233. Lubold, N., Walker, E., Pon-Barry, H.: Effects of voice-adaptation and social dialogue on perceptions of a robotic learning companion. In: Proceedings of the Human-Robot Interaction, pp. 1–8 (2016)

234. Luengo, I., Navas, E., Odriozola, I., Saratxaga, I., Hernaez, I., Sainz, I., Erro, D.: Modified LTSE-VAD algorithm for applications requiring reduced silence frame misclassification. In: Proceedings of the International Conference on Language Resources and Evaluation, LREC, pp. 1539–1544 (2010)

235. Maat, M.T., Truong, K.P., Heylen, D.: How turn-taking strategies influence users' impressions of an agent. In: Proceedings of the International Conference on Intelligent Virtual Agents (IVA), pp. 441–453. Springer, Berlin (2010)

236. Möller, S.: Perceptual quality dimensions of spoken dialogue systems: a review and new experimental results. In: Proceedings of the of Forum Acusticum, Budapest, p. 2681–2686 (2005)

237. Möller, S.: Quality of Telephone-Based Spoken Dialogue Systems. Springer, New York (2005)

238. Möller, S., Raake, A.: Motivation and introduction. In: Möller, S., Raake, A. (eds.) Quality of Experience: Advanced Concepts, Applications and Methods, pp. 3–10. Springer, Heidelberg (2014)

239. Möller, S., Raake, A. (eds.): Quality of Experience: Advanced Concepts, Applications and Methods. Springer, Heidelberg (2014)

240. Möller, S., Skowronek, J.: Quantifying the impact of system characteristics on perceived quality dimensions of a spoken dialogue service. In: Proceedings of the European Conference on Speech Communication and Technology, Geneva, vol. 3, pp. 1953–1956 (2003)

241. Möller, S., Engelbrecht, K.P., Schleicher, R.: Predicting the quality and usability of spoken dialogue services. Speech Commun. **50**, 730–744 (2009)

242. Mau, M.: Assessment of voice likability by means of psychological signals. Master's thesis, Technische Universität Berlin (2018)

243. Mayer, R.: Multimedia Learning, 2nd edn. University Press, Cambridge (2009)

244. McAleer, P., Todorov, A., Berlin, P.: How do you say 'hello'? personality impressions from brief novel voices. PLoS One **9**(3), e90779 (2014)

245. McCrae, R.R., Costa, P.T., Martin, T.A.: The NEO-PI-3: a more readable revised neo personality inventory. J. Pers. Assess. **84**, 261–270 (2005)

246. McCroskey, J., McCain, T.: The measurement of interpersonal attraction. Speech Monogr. **41**, 261–266 (1974)

247. McCroskey, J.C., Mehrley, R.S.: The effects of disorganization and nonfluency on attitude change and source credibility. Speech Monogr. **36**, 13–21 (1969)

248. McDonnell, R., Breidt, M., Bülthoff, H.: Render me real? Investigating the effect of render style on the perception of animated virtual humans. ACM Trans. Graph. **31**(4), 91, 1–11 (2012)

249. McGurk, H., MacDonald, J.: Hearing lips and seeing voices. Nature **264**(5588), 746–748 (1976)

250. Mc'Tear, M., Callejas, Z., Griol, D.: The Conversational Interface. Springer, Switzerland (2016)

251. Mead, R., Atrash, A., Matarić, M.J.: Automated proxemic feature extraction and behavior recognition: applications in human-robot interaction. Int. J. Soc. Robot. **5**, 367–378 (2013)

252. Mehrabian, A.: Some referents and measures of nonverbal behavior. Behav. Res. Methods Instrum. **1**, 213–217 (1969)
253. Mehrabian, A., Russell, J.: An Approach to Environmental Psychology. MIT Press, Cambridge (1974)
254. Mehu, M., Little, A.C., Dunbar, R.I.: Sex differences in the effect of smiling on social judgments: an evolutionary approach. J. Soc. Evol. Cult. Psychol. **2**, 103–121 (2008)
255. Mitchell, W.J.: The audio implicit association test: human preferences and implicit associations concerning machine voices. Dissertation, Indiana University (2009)
256. Moore, B.C.J.: An Introduction to the Psychology of Hearing, 5th edn. Academic Press, New York (2003)
257. Mumm, J., Mutlu, B.: Human-robot proxemics: Physical and psychological distancing in human-robot interaction. In: Proceedings of the Human-Robot Interaction, pp. 331–338 (2011)
258. Murry, T., Singh, S.: Multidimensional analysis of male and female voices. J. Acoust. Soc. Am. **68**, 1294–1300 (1980)
259. Murry, T., Singh, S., Sargent, M.: Multidimensional classification of abnormal voice qualities. J. Acoust. Soc. Am. **61**, 1630–1635 (1977)
260. NASA: NASA and Jamestown education module (2006). https://www.nasa.gov/pdf/166504main_Survival.pdf
261. Nass, C., Gong, L.: Maximized modality or constrained consistency? In: Auditory-Visual Speech Processing, pp. 1–5 (1999)
262. Nassiri, N., Powell, N., Moore, D.: Equilibrium theory revisited: mutual gaze and personal space in virtual environment. Virtual Reality **14**, 229–240 (2010)
263. Naumann, A., Hermann, F., Peissner, M., Henke, K.: Interaktion mit Informations- und Kommunikationstechnologie: Eine Klassifikation von Benutzertypen. In: Herczeg, M., Kindsmüller, M. (eds.) Mensch & Computer 2008: Viel Mehr Interaktion, pp. 37–45. Oldenbourg Wissenschaftsverlag, München (2008)
264. Naumann, A., Hermann, F., Niedermann, I., Peissner, M., Henke, K.: Interindividuelle Unterschiede in der Interaktion mit Informations- und Kommunikationstechnologie. In: Gross, T. (ed.) Mensch & Computer 2007, pp. 311–314. Oldenbourg Wissenschaftsverlag, München (2007)
265. Nawka, T., Anders, L., Cebulla, M., Zurakowski, D.: The speaker's formant in male voices. J. Voice **11**(4), 422–428 (1997)
266. Nenkova, A., Gravano, A., Hirschberg, J.: High frequency word entrainment in spoken dialogue. In: Proceedings of the 46th Annual Meeting of the Association for Computational Linguistics on Human Language Technologies, pp. 169–172, ACM, New York (2008)
267. Nolan, F., McDougall, K., Hudson, T.: Some acoustic correlates of perceived (dis)similarity between same-accent voices. In: Proceedings of the of the 17th International Congress of Phonetic Sciences, pp. 1506–1509 (2011)
268. Norton, R.W., Pettegrew, L.S.: Communicator style as an effect determinant of attraction. Commun. Res. **4**, 257–282 (1977)
269. Nosek, B.A., Smyth, F.L., Hansen, J.J., Devos, T., Lindner, N.M., Ranganath, K.A., Smith, C.T., Olson, K.R., Chugh, D., Greenwald, A.G., Banaji, M.R.: Pervasiveness and correlates of implicit attitudes and stereotypes. Eur. Rev. Soc. Psychol. **18**, 36–88 (2007)
270. Osgood, C.E., Suci, G., Tannenbaum, P.: The Measurement of Meaning. University of Illinois Press, Urbana, IL (1957)
271. Oviatt, S., Cohen, P.R.: The Paradigm Shift to Multimodality in Contemporary Computer Interfaces. Synthesis Lectures on Human-Centered Informatics. Morgan & Claypool, Wadsworth (2015)
272. Oviatt, S., Darves, C., Coulston, R.: Toward adaptive conversational interfaces: modeling speech convergence with animated personas. ACM Trans. Compu. Hum. Interact. **11**, 300–328 (2004)

273. Paeschke, A., Sendlmeier, W.F.: Die Reden von Rudolf Scharping und Oskar Lafontaine auf dem Parteitag der SPD im November 1995 in Mannheim—Ein sprechwissenschaftlicher und phonetischer Vergleich von Vortragsstilen. Z. Angew. Linguist. **27**, 5–39 (1997)

274. Paivio, A.: Mental Representations: A Dual Coding Approach. Oxford University Press, Oxford (1986)

275. Pardo, J.S.: On phonetic convergence during conversational interaction. J. Acoust. Soc. Am. **119**(4), 2382–2393 (2006)

276. Pérez, J., Gálvez, R., Gravano, A.: Disentrainment may be a positive thing: a novel measure of unsigned acoustic-prosodic synchrony, and its relation to speaker engagement. In: Proceedings of the Interspeech, pp. 1270–1274 (2016)

277. Pearl, C.: Designing Voice User Interfaces. O'Reilly, Beijing (2017)

278. Pentland, A.: Honest Signals: How They Shape Our World. MIT Press, Cambridge, MA (2008)

279. Perakakis, M., Potamianos, A.: Multimodal system evaluation using modality efficiency and synergy metrics. In: Proceedings of the International Conference on Multimodal Interaction (ICMI), pp. 9–16. ACM, New York (2008)

280. Peterson, R., Cannito, M., Brown, S.: An exploratory investigation of voice characteristics and selling effectiveness. J. Pers. Sell. Sales Manag. **15**(1), 1–15 (1995)

281. Petty, R.E., Wegener, D.T.: The elaboration likelihood model: Current status and controversies. In: Chaiken, S., Trope, Y. (eds.) Dual Process Theories in Social Psychology, pp. 41–72. Guilford Press, New York (1999)

282. Pfitzinger, H.R.: Local speech rate perception in German speech. In: Proceedings of the Congress of Phonetic Sciences (ICPhS), pp. 893–896 (1990)

283. Pickering, M.J., Garrod, S.: Toward a mechanistic psychology of dialogue. Behav. Brain Sci. **27**, 169–225 (2004)

284. Pickering, M.J., Garrod, S.: Alignment as the basis for successful communication. Res. Lang. Comput. **4**, 203–228 (2006)

285. Pinto-Coelho, L., Braga, D., Sales-Dias, M., Garcia-Mateo, C.: On the development of an automatic voice pleasantness classification and intensity estimation system. Comput. Speech Lang. **27**, 75–88 (2013)

286. Polychroniou, A.: The SSPNet—mobile corpus: from the detection of non-verbal cues to the inference of social behaviour during mobile phone conversations. Ph.D. thesis, University of Glasgow (2014)

287. Potamianos, G., Neti, C., Gravier, G., Garg, A., Senior, A.: Recent advances in the automatic recognition of audio-visual speech. Proc. IEEE **91**(9), 1306–1326 (2003)

288. Puckette, M.: The theory and technique of electronic music. http://puredata.info/ (2007)

289. Putnam, W.B., Street, R.L.J.: The conception and perception of noncontent speech performance: implications for speech-accommodation theory. Int. J. Sociol. Lang. **46**, 97–114 (1984)

290. R Core Team: R: A Language and Environment for Statistical Computing. R Foundation for Statistical Computing, Vienna, Austria (2018). https://www.R-project.org

291. Raake, A., Egger, S.: Quality and quality of experience. In: Möller, S., Raake, A. (eds.) Quality of Experience: Advanced Concepts, Applications and Methods, pp. 11–33. Springer, Heidelberg (2014)

292. Ramakrishna, A., Greer, T., Atkins, D., Narayanan, S.: Computational modeling of conversational humor in psychotherapy. In: Proceedings of the Interspeech (2018)

293. Rammstedt, B., John, O.P.: Measuring personality in one minute or less: a 10-item short version of the big five inventory in English and German. J. Res. Pers. **41**, 203–212 (2007)

294. Ray, G.: Vocally cued personality prototypes: An implicit personality theory approach. J. Commun. Monogr. **53**(3), 266–176 (1986)

295. Reeves, B., Nass, C.: The Eedia Equation: How People Treat Computers, Television, and New Media Like Real People and Places. Cambridge University Press, Cambridge (1996)

296. Reithinger, N., Klesen, M.: Dialog act classification using language models. In: Proceedings of the European Conference on Speech Communication and Technology, Rhodes, pp. 2235–2238 (1997)

297. Reitter, D., Moore, J.: Predicting success in dialogue. In: Proceedings of the Annual Meeting of the Association for Computational Linguistics (ACL), vol. 45, pp. 808–815 (2007)

298. Reitter, D., Moore, J.: Alignment and task success in spoken dialogue. J. Mem. Lang. **76**, 29–46 (2014)

299. Rieser, V., Lemon, O.: Reinforcement Learning for Adaptive Dialogue Systems. Springer, Berlin (2011)

300. Riether, N., Hegel, F., Wrede, B., Horstmann, G.: Social facilitation with social robots? In: Proceedings of the ACM/IEEE International Conference on Human-Robot Interaction, pp. 41–48. ACM, New York (2012)

301. Riggio, R., H.S., F.: Impression formation: the role of expressive behavior. J. Soc. Psychol. **50**, 421–427 (1986)

302. Rogers, I.: HCI Theory: Classical, Modern, and Contemporary. Synthesis Lectures on Human-Centered Informatics. Morgan & Claypool, Wadsworth (2012)

303. Roto, V., Law, E., Vermeeren, A., Hoonhout, J. (eds.): User experience whitepaper: bringing clarity to the concept of user experience (2011). www.allaboutux.org/uxwhitepaper. Result from Dagstuhl Seminar on Demarcating User Experience. Accessed 15–18 Sept 2010

304. Rule, N., Ambady, N.: First impressions of the face: Predicting success. Soc. Personal. Psychol. Compass **4**, 506–516 (2010)

305. Ruttkay, Z., C., D., Noot, H.: Embodied conversational agents on a common ground. a framework for design and evaluation. In: Ruttkay, Z., Pelachaud, C. (eds.) From Brows to Trust: Evaluating Embodied Conversational Agents, pp. 27–66. Springer, New York (2004)

306. Scapin, D., Senach, B., Trousse, B., Pallot, M.: User experience: buzzword or new paradigm? In: 5th International Conference on Advances in Computer-Human Interactions (ACHI), Valencia, pp. 336–341 (2012)

307. Schaller, M.: Evolutionary basis of first impressions. In: Ambady, N., Skowronski, J.J. (eds.) First Impressions, pp. 15–34. Guilford Press, New York (2008)

308. Schönbrodt, F., Back, M., Schmukle, S.: Tripler: an R package for social relations analyses based on round robin designs. Behav. Res. Methods **44**, 455–470 (2012)

309. Scherer, K.: Voice quality analysis of American and German speakers. J. Psycholinguist. Res. **3**, 281–298 (1974)

310. Scherer, K.: Personality inference from voice quality: the loud voice of extraversion. Eur. J. Soc. Psychol. **8**, 467–487 (1978)

311. Scherer, K.: Personality markers in speech. In: Scherer, K., Giles, H. (eds.) Social Markers in Speech, pp. 147–209. Cambridge University Press, Cambridge (1979)

312. Scherer, K., Giles, H.: Social Markers in Speech. Cambridge University Press, Cambridge (1979)

313. Schleicher, R.: Emotionen und Peripherphysiologie. Dissertation, Universität Köln, Lengerich (2009)

314. Schmitt, A., Minker, W.: Towards Adaptive Spoken Dialog Systems. Springer, New York (2013)

315. Schmitt, M., Bulterman, D.C.A., Cesar, P.S.: The contrast effect: QoE of mixed video-qualities at the same time. Quality User Experience **3**:7, 1–17 (2018)

316. Schoenenberg, K.: The quality of mediated-conversations under transmission delay. Ph.D. thesis, Technische Universität Berlin (2015)

317. Schuller, B., Batliner, A.: Computational Paralinguistics: Emotion, Affect and Personality in Speech and Language Processing. Wiley, London (2013)

318. Schuller, B., Steidl, S., Batliner, A., Nöth, E., Vinciarelli, A., Burkhardt, F., van Son, R., Weninger, F., Eyben, F., Bocklet, T., Mohammadi, G., Weiss, B.: Perceived speaker traits: personality, likability, pathology, and the first challenge. Comput. Speech Lang. **29**(1), 100–131 (2015)

319. Schulz v. Thun, F.: Miteinander reden: Störungen und Klärungen. Psychologie der zwischen-menschlichen Kommunikation. Rowohlt, Reinbek (1981)
320. Schweitzer, A., Lewandowski, N.: Convergence of articulation rate in spontaneous speech. In: Proceedings of the Interspeech, pp. 525–529 (2013)
321. Schweitzer, A., Walsh, M.: Exemplar dynamics in phonetic convergence of speech rate. In: Proceedings of the Interspeech, pp. 2100–2104 (2016)
322. Shepard, C.A., Giles, H., Le Poire, B.A.: Communication accommodation theory. In: Robinson, W.P., Giles, H. (eds.) The New Handbook of Language and Social Psychology, pp. 33–56. Wiley, New York (2001)
323. Shriberg, E., Bates, R., Stolcke, A., Taylor, P., Jurafsky, D., Ries, K., Coccaro, N., Martin, R., Meteer, M., Ess-Dykema, C.V.: Can prosody aid the automatic classification of dialog acts in conversational speech? Lang. Speech **41**, 439–487 (1998)
324. Silber-Varod, V., Lerner, A., Jokisch, O.: Automatic speaker's role classification with a bottom-up acoustic feature selection. In: Proceedings of the International Workshop on Grounding Language Understanding (GLU), pp. 52–56 (2017)
325. Singh, S., Murry, T.: Multidimensional classification of normal voice qualities. J. Acoust. Soc. Am. **64**(1), 81–87 (1978)
326. Skantze, G., Hjalmarsson, A.: Towards incremental speech generation in conversational systems. Comput. Speech Lang. **27**, 243–262 (2013)
327. Smith, B.L., Brown, B.L., Strong, W.J., Rencher, A.C.: Effects of speech rate on personality perception. Lang. Speech **18**, 145–152 (1975)
328. Spielberg, J.M., Stewart, J.L., Levin, R.L., Miller, G.A., Heller, W.: Prefrontal cortex, emotion, and approach/withdrawal motivation. Soc. Personal. Psychol. Compass **2**, 135–153 (2008)
329. Sproull, L., Subramani, M., Kiesler, S., Walker, J., Waters, K.: When the interface is a face. Hum. Comput. Interact. **11**, 97–124 (1996)
330. Stanton, C., Stevens, K.: Robot pressure: The impact of robot eye gaze and lifelike bodily movements upon decision-making and trust. In: Proceedings of the Social Robotics (ICSR). Lecture Notes in Artificial Intelligence, vol. 8755, pp. 330–339 (2014)
331. Stein, B., Stanford, T., Ramachandran, R., Perrault, T.J., Rowland, B.: Challenges in quantifying multisensory integration: alternative criteria, models, and inverse effectiveness. Exp. Brain Res. **198**, 113–126 (2009)
332. Steininger, S., Schiel, F., Rabold, S.: Annotation of multimodal data. In: Wahlster, W. (ed.) SmartKom: Foundations of Multimodal Dialogue Systems, Cognitive Technologies, pp. 571–596. Springer, Berlin (2006)
333. Stolcke, A., Coccaro, N., Bates, R., Taylor, P., Ess-Dykema, C.V., Ries, K., Shriberg, E., Jurafsky, D., Martin, R., Meteer, M.: Dialog act modeling for automatic tagging and recognition of conversational speech. Comput. Linguist. **26**, 339–373 (2000)
334. Street, R.L.: Evaluation of noncontent speech accommodation. Lang. Commun. **2**, 13–31 (1982)
335. Street, R.L.: Speech convergence and speech evaluation in fact-finding interviews. Hum. Commun. Res. **11**, 139–169 (1984)
336. Street, R.L.J.: Participant-observer differences in speech evaluation. J. Lang. Soc. Psychol. **4**, 125–130 (1985)
337. Street, R.L.J., Brady, R.M.: Speech rate acceptance ranges as a function of evaluative domain, listener speech rate and communication context. Commun. Monogr. **49**(4), 290–308 (1982)
338. Street, R.L.J., Brady, R.M., Putnam, W.B.: The influence of speech rate stereotypes and rate similarity or listeners' evaluation of speakers. J. Lang. Soc. Psychol. **2**(1), 37–56 (1983)
339. Su, P.H., Gasic, M., Young, S.: Reward estimation for dialogue policy optimisation. Comput. Speech Lang. **51**, 24–43 (2018)
340. Suhm, B., Waibel, A.: Toward better language models for spontaneous speech. In: Proceedings of the International Conference on Spoken Language Processing, Yokohama, pp. 831–834 (1994)

341. Sunnafrank, M., Ramirez, A.: At first sight: persistent relational effects of get-acquainted conversations. J. Soc. Pers. Relat. **21**, 361–379 (2004)

342. Susini, P., Lemaitre, G., McAdams, S.: Psychological measurement for sound description and evaluation. In: Berglund, B., Rossi, G., Townsend, J., Pendrill, L. (eds.) Measurement with Persons: Theory, Methods, and Implementation Areas, pp. 227–253. Psychology Press, New York (2012)

343. Takayama, L., Pantofaru, C.: Influences on proxemic behaviors in human-robot interaction. In: Proceedings of the International Conference on Intelligent Robots and Systems. IEEE, Piscataway (2009)

344. Takeuchi, A., Naito, T.: Situated facial displays: towards social interaction. In: Proceedings of the Conference on Human Factors in Computing Systems, pp. 450–455 (1995)

345. Tayal, S., Stone, S., Birkholz, P.: Towards the measurement of the actor's formant in female voices. In: Proceedings of the Elektronische Sprachsignalverarbeitung, pp. 286–293. TUD Press, Dresden (2017)

346. Thiran, J.P., Marqués, F., Bourlard, H.: Multimodal Signal Processing. Theory and Applications for Human-Computer Interaction. Academic Press, Oxford (2010)

347. Thomason, J., Nguyen, H.V., Litman, D.: Prosodic entrainment and tutoring dialogue success. In: Proceedings of the Artificial Intelligence in Education, pp. 750–753 (2013)

348. Titze, I.: Principles of Voice Production. Prentice Hall, Englewood Cliffs (1994)

349. Titze, I.R.: Acoustic interpretation of resonant voice. J. Voice **15**, 519–528 (2001)

350. Todorov, A., Pakrashi, M., Oosterhof, N.: Evaluating faces on trustworthiness after minimal time exposure. Soc. Cogn. **27**, 813–833 (2009)

351. Tractinsky, N., Cokhavi, A., Kirschenbaum, M., Sharfi, T.: Evaluating the consistency of immediate aesthetic perceptions of web pages. Int. J. Hum. Comput. Stud. **64**, 1071–1083 (2006)

352. Truong, K.P., Heylen, D.: Measuring prosodic alignment in cooperative task-based conversations. In: Proceedings of the Interspeech, pp. 843–846 (2012)

353. Tuch, A.N., Presslaber, E.E., Stöcklin, M., Opwis, K., Bargas-Avila, J.A.: The role of visual complexity and prototypicality regarding first impression of websites: working towards understanding aesthetic judgments. Int. J. Hum. Comput. Stud. **70**(11), 794–811 (2012)

354. Ultes, S., Budzianowski, P., Casanueva, I., Mrkšić, N., Rojas-Barahona, L., Su, P.H., Wen, T.H., Gašić, M., Young, S.: Domain-independent user satisfaction reward estimation for dialogue policy learning. In: Proceedings of the Interspeech, pp. 1721–1725 (2017)

355. van der Linden, D., Scholte, R.H., Cillessen, A.H., te Nijenhuis, J., Segers, E.: Classroom ratings of likeability and popularity are related to the Big Five and the general factor of personality. J. Res. Pers. **44**, 669–672 (2010)

356. van Dommelen, W., Moxness, B.: Acoustic parameters in speaker height and weight identification: sex-specific behaviour. Lang. Speech **38**, 267–287 (1995)

357. Vande Kamp, M.E.: Auditory implicit association tests. Dissertation, University of Washington (2002)

358. Venkatesh, V., Bala, H.: Technology acceptance model 3 and a research agenda on interventions. Decis. Sci. **39**, 273–315 (2008)

359. Venkatesh, V., Morris, M., Davis, G., Davis, F.: User acceptance of information technology: toward a unified vie. MIS Q. **27**, 425–278 (2003)

360. Vinciarelli, A., Salamin, H., Polychroniou, A., Mohammadi, G., Origlia, A.: From nonverbal cues to perception: personality and social attractiveness. In: Cognitive Behavioural Systems. Lecture Notes in Computer Science, vol. 7403, pp. 60–72. Springer, Berlin (2012)

361. Vogeley, K., Bente, G.: 'artificial humans': psychology and neuroscience perspectives on embodiment and nonverbal communication. Neural Netw. **23**, 1077–1090 (2010)

362. Voiers, W.D.: Perceptual bases of speaker identity. J. Acoust. Soc. Am. **36**, 1065–1073 (1964)

363. Von der Pütten, A., Krämer, N., Gratch, J.: Who's there? Can a virtual agent really elicit social presence? In: Proceedings of the PRESENCE, pp. 563–568 (2009)

364. Von der Pütten, A., Krämer, N., Gratch, J.: How our personality shapes our interactions with virtual characters—implications for research and development. In: Proceedings of the Intelligent Virtual Agents, pp. 208–221. Springer, Berlin (2010)

365. Wagner, P., Betz, S.: Speech synthesis evaluation: Realizing a social turn. In: Möbius, B., Steiner, I., Trouvain, J. (eds.) 28th Konferenz Elektronische Sprachsignalverarbeitung (ESSV), Saarbrücken, Studientexte zur Sprachkommunikation, pp. 167–173. TUD Press, Dresden (2017)

366. Walker, M.A., Passonneau, R.: DATE: a dialog act tagging scheme for evaluation of spoken dialog systems. In: Proceedings of the Human Language Technology Conference (HLT), pp. 1–8 (2001)

367. Walker, M.A., Kamm, C.A., Litman, D.J.: Towards developing general models of usability with PARADISE. Nat. Lang. Eng. **6**, 464–377 (2000)

368. Walker, M.A., Passonneau, R., Boland, J.E.: Quantitative and qualitative evaluation of Darpa Communicator spoken dialogue systems. In: Proceedings of the Annual Meeting on Association for Computational Linguistics, pp. 515–522 (2001)

369. Walker, M.A., Litman, D.J., Kamm, C.A., Abella, A.: PARADISE: a framework for evaluating spoken dialogue agents. In: Proceedings of the Association for Computational Linguistics, European Chapter (ACL/EACL), pp. 271–280 (1997)

370. Walker, M.A., Litman, D.J., Kamm, C.A., Abella, A.: Evaluating spoken dialogue agents with PARADISE: two case studies. Comput. Speech Lang. **12**, 317–347 (1998)

371. Wältermann, M.: Dimension-based Quality Modeling of Transmitted Speech. T-Labs Series in Telecommunication Services. Springer, Berlin (2013)

372. Walton, T., Evans, M.: The role of human influence factors on overall listening experience. Quality User Experience **3:1**, 1–16 (2018)

373. Ward, A., Litman, D.: Dialog convergence and learning. In: Proceedings of the Artificial Intelligence in Education, pp. 1–8 (2007)

374. Ward, N., Nakagawa, S.: Automatic user-adaptive speaking rate selection for information delivery. In: Proceedings of the 7th International Conference on Spoken Language Processing (ICSLP), pp. 549–552 (1990)

375. Ward, N.G., Abu, S.: Action-coordinating prosody. In: Speech Prosody, pp. 629–633 (2016)

376. Wechsung, I., De Moor, K.: Quality of Experience versus User Experience. In: Möller, S., Raake, A. (eds.) Quality of Experience: Advanced Concepts, Applications and Methods, pp. 35–54. Springer, Heidelberg (2014)

377. Wechsung, I., P., E., Schleicher, R., Möller, S.: Investigating the social facilitation effect in human-robot-interaction. In: International Workshop on Spoken Dialogue Systems Technology (IWSDS), pp. 1–10 (2012)

378. Wechsung, I., Ehrenbrink, P., Schleicher, R., Möller, S.: Investigating the social facilitation effect in human-robot interaction. In: International Workshop on Spoken Dialogue Systems (IWSDS), pp. 125–134 (2012)

379. Wechsung, I., Weiss, B., Ehrenbrink, P., Möller, S.: Development and validation of the conversational agents scale (CAS). In: Interspeech, Lyon, pp. 1106–1110 (2013)

380. Wechsung, I., Engelbrecht, K.P., Kühnel, C., Möller, S., Weiss, B.: Measuring the quality of service and quality of experience of multimodal human-machine interaction. J. Multimodal User Interfaces **6**(1), 73–85 (2012)

381. Wechsung, I., Schulz, M., Engelbrecht, K.P., Niemann, J., Möller, S.: All users are (not) equal—the influence of user characteristics on perceived quality, modality choice and performance. In: Workshop on Paralinguistic Information and its Integration in Spoken Dialogue Systems (IWSDS), pp. 175–188 (2011)

382. Weirich, M.: Die attraktive Stimme: Vocal Stereotypes. Eine phonetische Analyse anhand akustischer und auditiver Parameter. Verlag Dr. Müller, Saarbrücken (2010)

383. Weiss, B.: Rate dependent spectral reduction for voiceless fricatives. In: 9th Interspeech, Brisbane, p. 1968 (2008)

384. Weiss, B.: Prosodische elemente vokaler sympathie. In: Wagner, P. (ed.) 24th Konferenz Elektronische Sprachsignalverarbeitung (ESSV), Bielefeld. Studientexte zur Sprachkommunikation, vol. 65, pp. 212–217. TUD Press, Dresden (2013)
385. Weiss, B.: Akustische Korrelate von Sympathieurteilen bei Hörern gleichen Geschlechts. In: 26th Konferenz Elektronische Sprachsignalverarbeitung (ESSV), Eichstädt. Studientexte zur Sprachkommunikation, vol. 78, pp. 165–171. TUD Press, Dresden (2015)
386. Weiss, B.: Voice descriptions by non-experts: validation of a questionnaire. In: Proceedings of the Phonetics & Phonology, pp. 228–231 (2016)
387. Weiss, B., Burkhardt, F.: Voice attributes affecting likability perception. In: 11th Interspeech, Makuhari, pp. 1934–1937 (2010)
388. Weiss, B., Burkhardt, F.: Is 'not bad' good enough? Aspects of unknown voices' likability. In: 13th Interspeech, Portland, pp. 1–4 (2012)
389. Weiss, B., Hillmann, S.: Feedback matters: applying dialog act annotation to study social attractiveness in three-party conversations. In: ACL-ISO Workshop on Interoperable Semantic Annotation, Portorož, pp. 55–58 (2016)
390. Weiss, B., Möller, S.: Wahrnehmungsdimensionen von Stimme und Sprechweise. In: Kröger, B., Birkholz, P. (eds.) 22th Konferenz Elektronische Sprachsignalverarbeitung (ESSV), Aachen. Studientexte zur Sprachkommunikation, vol. 61, pp. 261–268. TUD Press, Dresden (2011)
391. Weiss, B., Schoenenberg, K.: Conversational structures affecting auditory likeability. In: Interspeech, pp. 1791–1795 (2014)
392. Weiss, B., Tönges, R.: Automatic adaption of spoken dialog systems for public and working environments. In: IADIS International Conference on Interfaces and Human Computer Interaction (IHCI), Lisbon, pp. 284–288 (2012)
393. Weiss, B., Burkhardt, F., Geier, M.: Towards perceptual dimensions of speakers' voices: Eliciting individual descriptions. In: Proceedings of the Workshop on Affective Social Speech Signals, Grenoble (2013)
394. Weiss, B., Estival, D., Stiefelhagen, U.: Studying vocal perceptual dimension of non-experts by assigning overall speaker (dis-)similarities. Acta Acust. Acust. **104**, 174–184 (2018)
395. Weiss, B., Hillmann, S., Michael, T.: Kontinuierliche Schätzung von Sprechgeschwindigkeit mit einem Rekurrenten Neuronalen Netzwerk. In: Berton, A., Haiber, U., Minker, W. (eds.) 29th Konferenz Elektronische Sprachsignalverarbeitung (ESSV), Ulm, Studientexte zur Sprachkommunikation, pp. 186–191. TUD Press, Dresden (2018)
396. Weiss, B., Möller, S., Polzehl, T.: Zur Wirkung menschlicher Stimme auf die wahrgenommene Sympathie—Einfluss der Stimmanregung. In: Mixdorff, H. (ed.) 21th Konferenz Elektronische Sprachsignalverarbeitung (ESSV), Berlin. Studientexte zur Sprachkommunikation, vol. 58, pp. 56–63. TUD Press, Dresden (2010)
397. Weiss, B., Möller, S., Schulz, M.: User differences in evaluating multimodal HCI. In: 5th International Conference on Advances in Computer-Human Interactions (ACHI), Valencia, pp. 354–359 (2012)
398. Weiss, B., Trouvain, J., Burkhardt, F.: Acoustic correlates of likable speakers in the NSC database. In: Barkat-Defradas, M., Weiss, B., Trouvain, J., Ohala, J.J. (eds.) Chapter 12: Voice Attractiveness: Studies on Sexy, Likable, and Charismatic Speakers, Prosody, Phonology and Phonetics, pp. 99–115. Springer, Berlin (2019)
399. Weiss, B., Wechsung, I., Hillmann, S., Möller, S.: Multimodal HCI: exploratory studies on effects of first impression and single modality ratings in retrospective evaluation. J. Multimodal User Interfaces **11**(2), 115–131 (2017)
400. Weiss, B., Wechsung, I., Marquardt, S.: Assessing ICT user groups. In: ACM NordiCHI, Copenhagen, pp. 275–283 (2012)
401. Weiss, B., Willkomm, S., Möller, S.: Evaluating an adaptive dialog system for the public. In: Interspeech, Lyon, pp. 2034–2038 (2013)
402. Weiss, B., Wechsung, I., Kühnel, C., Möller, S.: Evaluating embodied conversational agents in multimodal interfaces. Comput. Cogn. Sci. **1:6**, 1–21 (2015)

403. Weiss, B., Hacker, A., Moshona, C., Rudawski, F., Ruhland, M.: Studying vocal social attractiveness by re-synthesis—results from two student projects applying acoustic morphing with tandem-straight. In: Trouvain, J., Steiner, I., Möbius, B. (eds.) 28th Konferenz Elektronische Sprachsignalverarbeitung (ESSV), Saarbrücken, Studientexte zur Sprachkommunikation, pp. 316–323. TUD Press, Dresden (2017)

404. Weiss, B., Kühnel, C., Wechsung, I., Fagel, S., Möller, S.: Quality of talking heads in different interaction and media contexts. Speech Commun. **52**(6), 481–492 (2010)

405. Weiss, B., Kühnel, C., Wechsung, I., Möller, S., Fagel, S.: Web-based evaluation of talking heads: how valid is it? In: 9th International Conference on Intelligent Virtual Agents (IVA), Amsterdam, LNAI 5773, p. 552. Springer, Berlin (2009)

406. Weiss, B., Guse, D., Möller, S., Raake, A., Borowiak, A., Reiter, U.: Temporal development of quality of experience. In: Möller, S., Raake, A. (eds.) Quality of Experience: Advanced Concepts, Applications and Methods, pp. 133–147. Springer, Heidelberg (2014)

407. Wiggins, J.S., Trapnell, P., Phillips, N.: Psychometric and geometric characteristics of the revised interpersonal adjective scales (IAS-R). Multivar. Behav. Res. **23**(4), 517–530 (1988)

408. Williams, K.D., Cheung, C.K.T., Choi, W.: Cyberostracism: Effects of being ignored over the internet. J. Pers. Soc. Psychol. **79**, 748–762 (2000)

409. Winkler, R.: Merkmale junger und alter Stimmen—Analyse ausgewählter Parameter im Kontext von Wahrnehmung und Klassifikation. Dissertation, Technische Universität Berlin, Berlin (2009)

410. Włodarczak, M., Simko, J., Wagner, P.: Temporal entrainment in overlapped speech: cross-linguistic study. In: Proceedings of the Interspeech, pp. 615–618 (2012)

411. Wolters, M., Georgila, K., MacPherson, S., Moore, J.: Being old doesn't mean acting old: older users' interaction with spoken dialogue systems. ACM Trans. Accessible Comput. **2**(1), 1–39 (2009)

412. Wortman, J., Wood, D.: The personality traits of liked people. J. Res. Pers. **45**, 519–528 (2011)

413. Wright Hastie, H., Poesio, M., Isard, S.: Automatically predicting dialoguestructure using prosodic features. Speech Commun. **36**, 63–79 (1998)

414. Wu, Z., Swietojanski, P., Veaux, C., Renals, S., King, S.: A study of speaker adaptation for DNN-based speech synthesis. In: Proceedings of the Interspeech, pp. 879–883 (2015)

415. Yang, Z., Narayanan, S.: Analyzing temporal dynamics of dyadic synchrony in affective interactions. In: Proceedings of the Interspeech, pp. 42–46 (2016)

416. Yee, N., Bailenson, J., Rickertsen, K.: A meta-analysis of the impact of the inclusion and realism of human-like faces on user experiences in interfaces. In: Proceedings of the Conference on Human Factors in Computing Systems, pp. 1–10 (2007)

417. Zuckermann, M., Miyake, K.: The attractive voice: what makes it so? J. Nonverbal Behav. **17**, 119–135 (1993)

418. Zuckermann, M., Miyake, K.: Beyond personality impressions: effects of physical and vocal attractiveness on false consensus, social comparison, affiliation and assumed and perceived similarity. J. Pers. **61**, 411–437 (1993)

419. Zwicker, E., Fastl, H.: Psychoacoustics: Facts and Models, 3rd edn. Springer, Berlin (2007)

Index

© Springer Nature Switzerland AG 2020

B. Weiss, *Talker Quality in Human and Machine Interaction*, T-Labs Series in Telecommunication Services, https://doi.org/10.1007/978-3-030-22769-2

Printed in the United States
By Bookmasters